Introduction to Bioinformatics and Microarray Technology

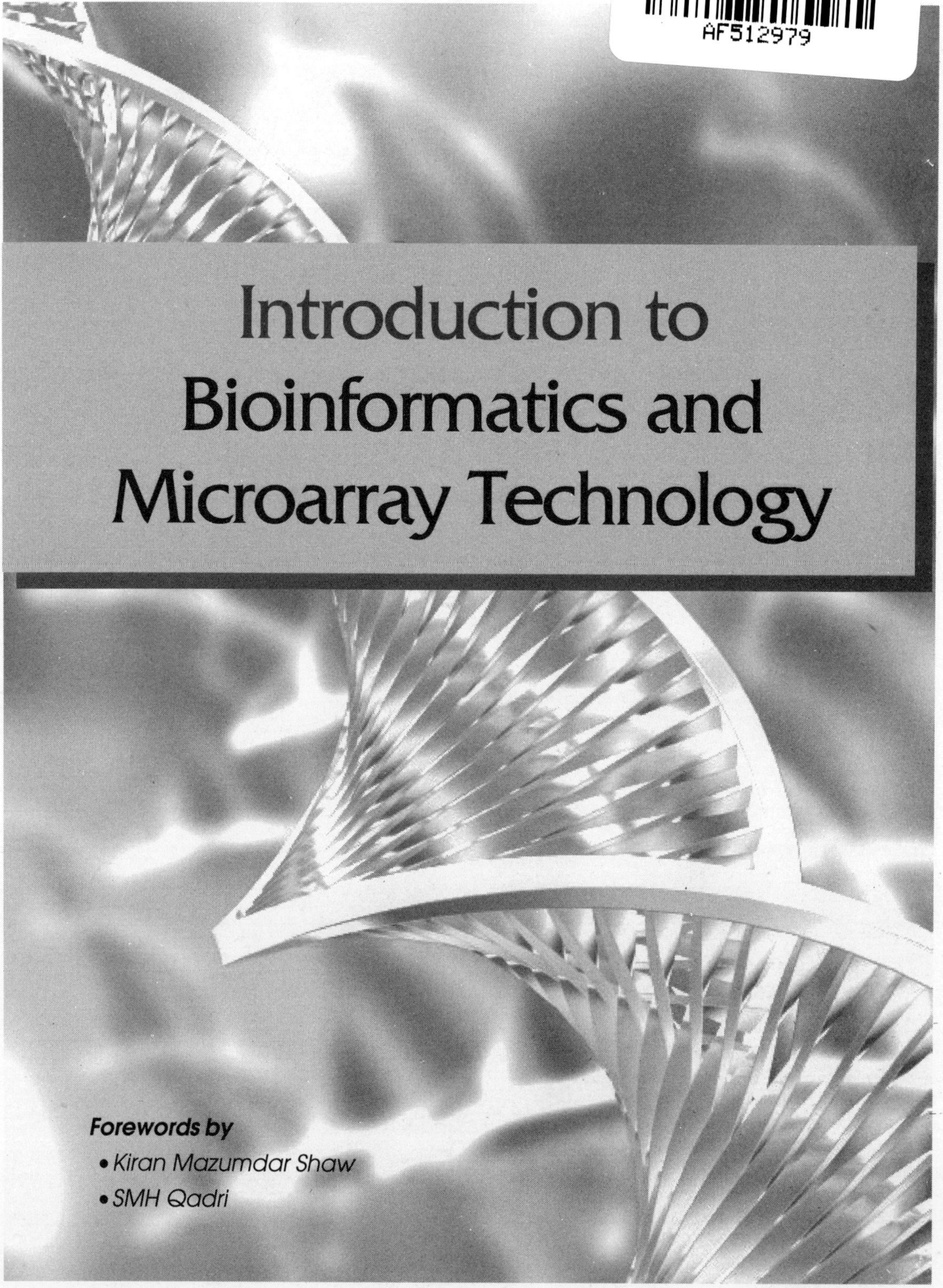

Forewords by

- *Kiran Mazumdar Shaw*
- *SMH Qadri*

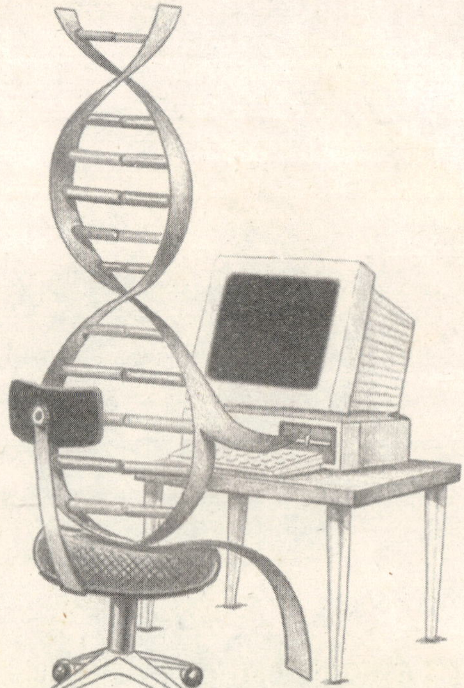

Introduction to Bioinformatics and Microarray Technology

Abhilash M. BEng, MTech, MISTE

Senior Lecturer
Department of Biotechnology
The Oxford College of Fngineering
Bangalore, India

Editor

Asian Journal of Experimental Biological Sciences
Journal of Bioinformatics and Sequence Analysis
Internet Journal of Bioengineering

Reviewer

International Journal of Biometrics and Bioinformatics
Indian Journal of Science and Technology
The Journal of Scientific Research
Avicenna Journal of Medical Biotechnology
Journal of AIDS and HIV Research
Clinical Reviews and Opinion
International Journal of Pharmacy and Biosciences
Journal of Developmental Biology and Tissue Engineering

CBS

CBS Publishers & Distributors Pvt Ltd

New Delhi • Bengaluru • Pune • Kochi • Chennai

Disclaimer

Science and technology are constantly changing fields. New research and experience broaden the scope of information and knowledge. The author has tried his best in giving information available to him while preparing the material for this book. Although, all efforts have been made to ensure optimum accuracy of the material, yet it is quite possible some errors might have been left uncorrected. The publisher, printer and the author will not be held responsible for any inadvertent errors or inaccuracies.

Introduction to Bioinformatics and Microarray Technology

ISBN: 978-81-239-1878-5

First Edition: 2010

Published by Satish Kumar Jain and produced by Vinod K. Jain for
CBS Publishers & Distributors Pvt Ltd
4819/XI Prahlad Street, 24 Ansari Road, Daryaganj,
New Delhi 110 002, India.
Ph: 23289259, 23266861/67 Fax: +91-011-23243014
Website: www.cbspd.com
e-mail: delhi@cbspd.com; cbspubs@vsnl.com; cbspubs@airtelmail.in.

Branches

- Bengaluru: Seema House 2975, 17th Cross, K.R. Road, Banasankari 2nd Stage, Bengaluru 560 070, Karnataka
  Ph: +91-80-26771678/79 Fax: +91-80-26771680 e-mail: bangalore@cbspd.com
- Pune: Shaan Brahmha Complex, 631/632 Basement, Appa Balwant Chowk, Budhwar Peth, next to Ratan Talkies, Pune 411 002, Maharashtra
  Ph: +91-20-24464057/58 Fax: +91-20-24464059 e-mail: pune@cbspd.com
- Kochi: 36/14 Kalluvilakam, Lissie Hospital Road, Kochi 682 018, Kerala
  Ph: +91-484-4059061-65 Fax: +91-484-4059065 e-mail: cochin@cbspd.com
- Chennai: 20, West Park Road, Shenoy Nagar, Chennai 600 030, Tamilnadu
  Ph: +91-44-26260666, 26208620 Fax: +91-44-45530020 email: chennai@cbspd.com

Printed at India Binding House, Noida (UP)

my unconditional love

and dedication

to

my father Mr Manohar

mother Mrs Geetha

grandfather Mr Nagesh Rao

and

sister Ms Anusha

Foreword

A disease-free world is the aspiration of every citizen of the world. Preventative and prognostic medicine is what can help realize this vision where bioinformatics and microarray technology are at the fulcrum of this endeavour.

Rapid advances in biological and molecular research are expanding our understanding of life itself. They are helping us understand, prevent and treat previously incurable genetically triggered diseases. Personalized medicine, molecular diagnostics and the synthesis of therapeutics and diagnostics are the promises bioinformatics offers by combining information technology with molecular biology. Its application is not merely restricted to human health but spans fields like environment, agriculture and biotechnology, positively impacting all facets of life.

The Human Genome Project has spawned much interest in the DNA sequence of the human genome. Although this has led to the identification of a large number of new genes within sequences, organizing this vast amount of information so as to make it easy to understand and use is a challenge that researchers are still grappling with. Earlier, conventional methods of assaying gene expression allowed researchers to painstakingly survey a few genes at a time. However, today one single experiment with microarrays is all it takes for scientists to understand the expression of several genes. Such breakthroughs in methodology have allowed researchers to quickly plot the genetic causes of many diseases.

Today, the world is keenly following new developments in bioinformatics and microarray technology and their applications with numerous research journals and scientific books tracking the latest trends. Yet, such literature does not help researchers working in life sciences, computer science and information technology to build their own customized applications. Introduction to Bioinformatics and Microarray Technology by Mr Abhilash M., seeks to bridge this gap by highlighting the potential of computation in biological research and introducing existing databases, tools and resources necessary for such research.

The chapter on the role of IT in bioinformatics is especially important. Though we have a large number of software professionals who understand the importance of bioinformatics, it is necessary for them to be familiar with the basics of biology, if they are to contribute productively to this field. In the same vein, biologists may not fully comprehend the potential of bioinformatics in their research. This book caters to the needs of both IT professionals and biologists who want to use bioinformatics and microarray technology in their research. That it helps researchers gain expertise in making use of biological databases, tools and bioinformatics resources, is an added advantage of the book.

Presently, books on microarray technology available in the market deal with advanced techniques. Introduction to Bioinformatics and Microarray Technology is the first bioinformatics book which deals with the basic as well as advanced techniques in the fabrication of microarrays.

Students pursuing graduate and postgraduate studies in genetics, chemistry, molecular biology, microbiology, veterinary science, pharmacy, mathematics, physics as well as IT professionals, engineers and medical practitioners will find this book useful. An exhaustive guide to understanding the basic concepts of genome sequence analysis and bioinformatics, its lucid contents cover all key topics on bioinformatics and microarray technology. Mr Abhilash has done well to keep in mind the syllabi of Indian as well as global universities on the subject.

With more than 30 scientific articles published in national and international journals, and with years of experience teaching biotechnology, Mr Abhilash is himself an avid researcher. He is involved in numerous collaborative projects that include molecular modelling, in silico drug design, molecular simulation and dynamics. He is an editor and reviewer of several leading publications in this field.

While bioinformatics and microarray technology are emerging fields, I am certain they will help realize the global dream of a healthy world that is devoid of diseases. However, it is armies of skilled and talented researchers and workers who will finally exploit the potential that bioinformatics and microarray technology offers. I am sure this book will not merely educate researchers but also spark interest in this field in the hearts and minds of budding scientists.

Kiran Mazumdar Shaw
Chairperson and Managing Director
Biocon Ltd

Foreword

It is widely recognized that the field of biology is in the midst of a 'data explosion'. A series of technical advances in recent years has increased the amount of data that biologists can record about different aspects of an organism at the genomic, transcriptomic and proteomic levels. This data is, of course, vital to advancing our knowledge. In recent years, the discipline of bioinformatics has allowed biologists to make full use of the advances in computer science and computational statistics in analysing this data. Massive data acquisition technologies, such as genome sequencing, high throughput drug screening, and DNA arrays are in the process of revolutionizing biology and medicine. Using the mRNA of a given cell, at a given time, under a given set of conditions, DNA microarrays can provide a snapshot of the level of expression of all the genes in the cell. Such snapshots can be used to study fundamental biological phenomena such as development or evolution, to determine the function of new genes, to infer the role that individual genes or group of genes may play in diseases, and to monitor the effect of drugs and other compounds on gene expression

This interdisciplinary book entitled Introduction to Bioinformatics and Microarray Technology will be an essential reading for researchers wanting to take advantage of this powerful new technology. This book has been written to engage and interest readers from many disciplines. For biologists there is a full introduction to the challenges for computer science and computer scientists should also find the chapters on biology and bioinformatics informative. Practising bioinformaticians are also likely to find the book enlightening and a collection such as this provides a single resource for many intelligent problem-solving techniques in bioinformatics.

I strongly feel that the efforts put in by Mr Abhilash will help to serve as a handy and valuable source of information to meet the tremendously important challenge of training the 21st century computational biologists. He has undertaken the difficult task of organizing the knowledge in this field in a logical progression and presenting it in a digestible form and has done an excellent job. I hope this book will certainly bring an impact on biological research and, in turn, on progress in bioinformatics and microarray technology.

Dr SMH Qadri
Director and Coordinator, DBT Bioinformatics Centre
Central Sericultural Research and Training Institute
(Central Silk Board, Ministry of Textiles, Government of India)
Mysore, India

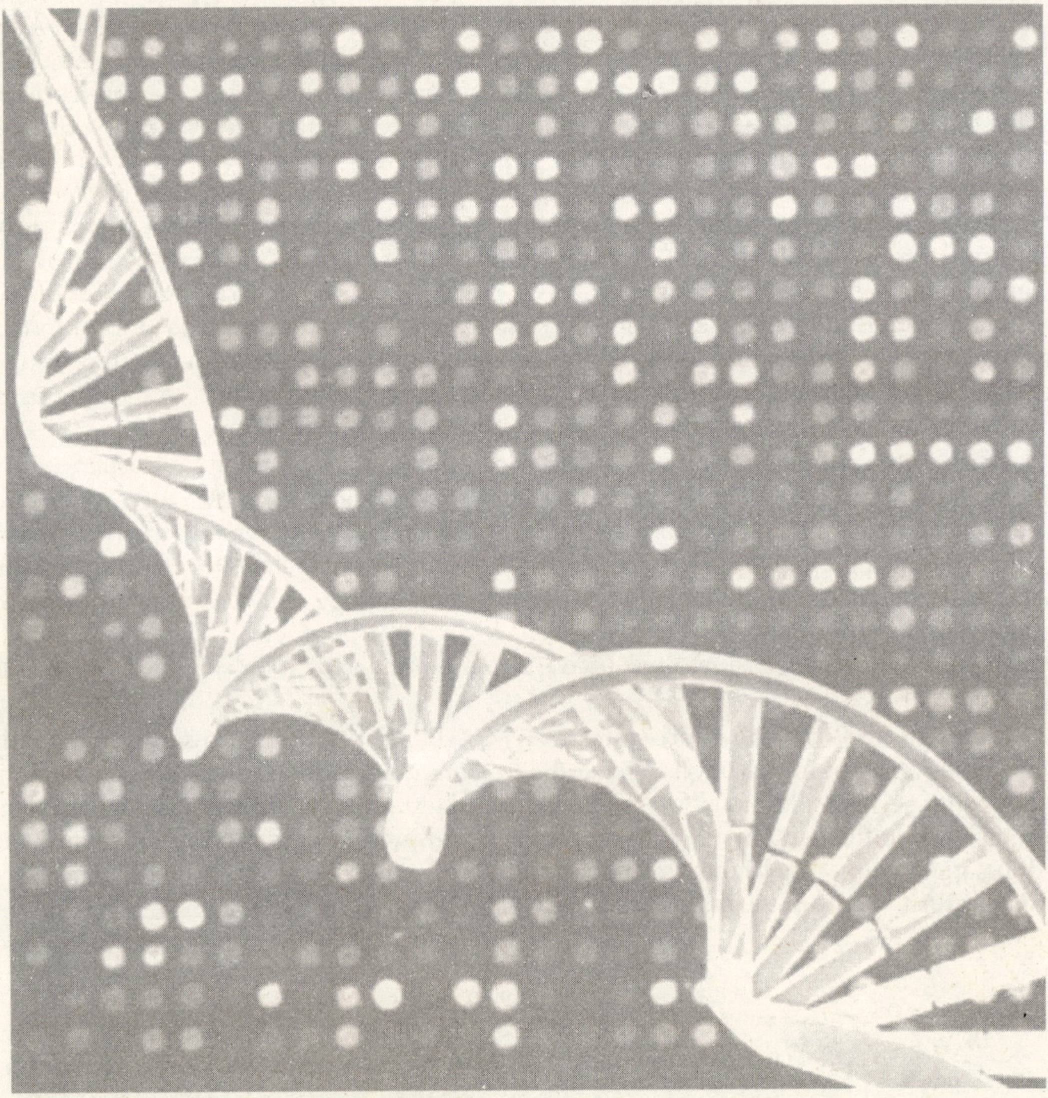

Preface

Bioinformatics has emerged as a strategic discipline at the frontier between biology and computer science, impacting medicine, biotechnology, and society in many ways. It is emerging as an ever-evolving new branch of science in which computer tools are applied to collect, store, and analyze biological data to generate new biological information. Over the past few years, major progress in the field of molecular biology, coupled with rapid advances in genomic technologies, has led to an explosive growth in the biological information distributed in a variety of biological databases.

The tsunami of biological information available upon completion of various genome projects and their methodical organization using personal computer has lead to birth of field of bioinformatics. Large databases of biological information create both challenging data mining problems and opportunities, each requiring new ideas. The large amounts of data create a critical need for theoretical, algorithmic and software advances in storing, retrieving, networking, processing, analyzing, navigating, and visualizing biological information.

Most biological science students and researchers are starting to use computers as more than word-processing or data-collection and plotting devices. Many do not have background in computer science or computational theory, and to them, the field of bioinformatics may seem hopelessly large and complex. This book, motivated by my interactions with my students and colleagues, is by no means a comprehensive reference book on all important aspects of bioinformatics.

Bioinformatics and microarray technology have become major fields with hugely important applications. Although professional journals and books throughout the world provide access to new details on bioinformatics and microarray technology. It is still very hard for practitioners in the field of life sciences, computer science and information technology to build their own tailor-made applications.

One of my motivations for writing this book was to explain the potential of computation in field of biology and also to introduce biologists about existing databases, tools and resources which will be very useful in their research. Microarray and biochip is a multiplex technology used in molecular biology and in medicine. It is important for biologists to know the potential of this field.

This book is expected to serve as a useful reference for researchers and students embarking on the exciting field of bioinformatics, drug discovery and microarray technology. Please feel-free to share your comments, suggestions for improvement of this book by writing to me at abhibiotek@gmail.com.

Abhilash M.

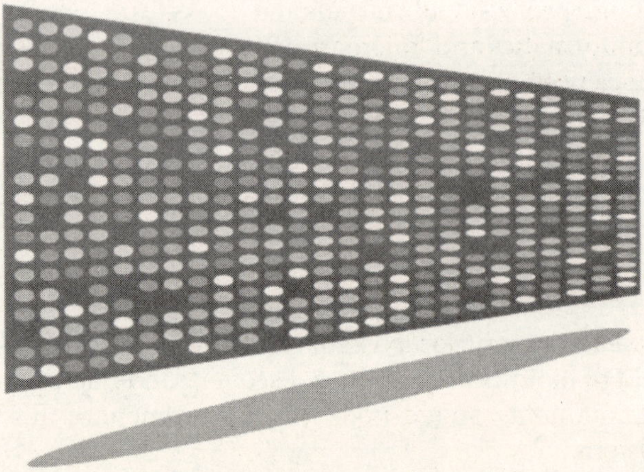

Acknowledgments

The realization of this piece of work would not have been possible without the contributions of many persons who rendered their help in one way or the other.

I would like to thank Mr Satish Kumar Jain, Mr YN Arjuna, Mr Deepak Rao, Mr Pradeep Kumar and everyone at CBS Publishers & Distributors Pvt Ltd for their skill, enthusiasm, support, patience and excellent professional approach in publishing this book.

I express my profound gratitude and thanks to Mr S Narasaraju, the founder chairman, and Mr SNVL Narasimha Raju, the executive director of The Oxford Educational Institutions for their cordial encouragement.

I extend my thanks to Principal Dr T Krishnan and Vice Principal Dr Nagaraj R. of The Oxford College of Engineering for their encouragement and support in writing this book.

I wish to record my sincere thanks to Dr KN Somasekharan, former Dean, SCBT Sastra University, and Prof K Basavaraju, former Director of Technical Education, Government of Karnataka, for their valuable suggestions and moral support in writing this book.

My special thanks to Mr Kumaran M, chief librarian of The Oxford College of Engineering, for his constant support, encouragement and motivation for writing this book.

My heartful appreciation to Dr Kusum Paul, Dr BK Manjunath, Mr Divakara, Mr Ganesh and all my colleagues at The Oxford College of Engineering, Bangalore, for their steady assistance and constant feedback extended in writing this book.

Finally, I would like to thank all my relatives and my best friends Mr Krishna, Mr Harshith, Ms Kavitha, Mr Raja and Mr Sinosh for their unceasing moral support and affection.

Abhilash M.

Contents

1

Introduction to Bioinformatics

What is Bioinformatics?

Bioinformatics is the application of computers in biological sciences and especially analysis of biological sequence data. It is concerned with capturing, storing, graphically displaying, modelling and ultimately distributing biological information. It is becoming an essential tool in molecular biology as genome projects generate vast quantities of data. With new sequences being added to DNA databases on an average, once every minute, there is a pressing need to convert this information into biochemical and biophysical knowledge by deciphering the structural, functional and evolutionary clues encoded in the language of biological sequences.

What bioinformatics therefore offers to the researcher, the entrepreneur, or the venture capitalist is an enormous and exciting array of opportunities to discover how living systems metabolise, grow, combat disease, reproduce and regenerate. The current knowledge represents only the tip of the iceberg. Exciting and startling discoveries are being made everyday through bioinformatics, which is building up an extensive encyclopedia from which life's mysteries will be unraveled. The importance of computational science in collating this information and its simultaneous interpretation by biologists is the underlying ethos of bioinformatics.

Having an interest in biology and having a strong inclination towards genetics is all right. But the most important thing is that, biocomputing requires lots of software professionals. And there is more to do for these people than the experts in biology.

Current Knowledge and the 'Central Dogma'

What has been described so far is the 'central dogma' in biology: that one gene by and large produces one mRNA that by and large produces one protein. More specifically, the central dogma consists of six 'axioms'.

1. DNA replicates its information in a process that involves many replication enzymes.
2. DNA codes for the production of messenger RNA (mRNA) during transcription.
3. In eukaryotic cells, the mRNA is processed (essentially by splicing) and migrates from the nucleus to the cytoplasm.
4. mRNA carries coded information to ribosomes, where protein is synthesized using the mRNA during translation.
5. Proteins do not code for the production of proteins, RNA or DNA.
6. Proteins are involved in almost all biological activities, structural or enzymatic.

In fact, transcription and translation are even more complicated than previously described. Some gene products are not translated at all but function in their RNA form after transcription. For instance, genes that code for tRNA and ribosomes cannot be translated. If they could, they would depend on tRNA and ribosomes for the translation! Instead, after transcription, these RNA molecules exit the nucleus and perform their roles in the translation of normally transcribed and translated genes. This has led to the identification of several different types of RNA, the most common of which are messenger RNA (mRNA), transfer RNA (tRNA), ribosomal RNA (rRNA) for the building of the ribosomes, and small nuclear RNA (snRNA) that help edit the mRNA. mRNA is the primary messenger synthesized from a gene segment of DNA and carrying the code into the cytoplasm where protein synthesis occurs. rRNA (ribosomal) in the cytoplasm and protein combine to form a nucleoprotein (ribosome) that serves as the translation site and carries the enzymes necessary for protein synthesis. Several ribosomes may be attached to a single mRNA at any time. tRNA (transfer) contains about 75 nucleotides, three of which form a tRNA anticodon, and one amino acid. The tRNA reads the code and carries the amino acid to be incorporated into the developing protein.

Also, there is growing evidence that, for many genes, there are many more 'splice variants' than previously believed, where different combinations of introns and exons exist. It is not a simple matter of introns being removed and the remaining exons being sent sequentially to the ribosomes. One gene can produce many different types of polypeptide chain depending on how many introns are removed and how many the exons are shuffled or differently spliced. This means that, while the human genome may consist of only 30000 genes, the genes may produce many more proteins. It is not currently clear what the relationship between gene and protein is in humans, or how many human genes can be alternatively spliced, but estimates vary from between 1:10 to 1:100 for the gene: protein ratio and from 60 % up to 75 % of human genes having alternative splice variants. That is, although humans have only 30000 genes, these may well produce anything between 200000 and two million proteins through alternative splicing. Interest in splice variants has led to the conjecture that, in addition to the genome (total set of DNA and genes) and proteome (total set of proteins), it is in the spliceosome (total set of alternative splice variants for a particular gene) that the real answers will be found as to how genes are mapped on to proteins and how many proteins are actually capable of being made.

DNA polymerase which replicates chromosomes is so accurate that there is only one error in every 10^7 nucleotide pairs. A DNA mismatch repair system also operates to correct nearly all of these errors, increasing the overall accuracy to one error in 10^9 nucleotides copied. While thousands of random chemical changes are created every day in human DNA, the vast majority are eliminated by DNA repair. For instance, some varieties of yeast have about 50 genes devoted to gene repair. It is not known how many human genes are devoted to gene repair.

Why Proteins are Important?

As structural molecules, they provide much of the cytoskeletal framework of cells and also help cells form tissue. Proteins also carry signals from one part of the body to another, or from one cell to another. Proteins can also act as a transport system, carrying molecules such as oxygen in the circulatory system so that all cells can have access to this important element. The human immune system is also dependent on proteins for detecting the arrival of pathogens as they enter the human system and for helping to mount an effective immune system defence. As enzymes, proteins act as biological catalysts that speed up the rate of cellular reactions. The chemical composition of one cell could be placed in a test tube and observed. After some time, some chemical reactions naturally occurring in the test tube might be noted. There will be a long delay because the activation energy required to start a chemical reaction acts like an energy barrier over which the molecules must be raised for a reaction to take place. An enzyme effectively lowers the activation energy required for a reaction to proceed. An enzyme locks on to a molecule, starts a reaction, and then is released unchanged. The rate of enzyme combination and release is called the turnover rate and is about 1000 times a second for most enzymes, with variation between 100/sec and 10 million/sec. The increase in reaction rate achieved by enzymes ranges from a minimum of about a million to as much as a trillion times faster than an uncatalysed reaction at equivalent concentrations and temperatures.

However, enzymes cannot work unless they have folded in the right way. That is, it is the structure of the enzyme that determines what it does and whether it does it. If enzymes do not fold in the right way, they cannot carry out their enzymatic activity, because they will not be able to lock on to their target molecules (which also have complex three-dimensional structures) to start the appropriate reaction. The study of protein folding and protein structure is a key element in understanding what proteins do. However, what is actually observed in organisms are already-folded proteins. Trying to 'straighten' a protein so that the sequence of amino acids is revealed is a major problem in proteomics. The number of well understood, sequenced and structured proteins is small compared with the total number of proteins that exist. Also, as will be seen later, trying to determine the structure of a protein from its DNA, mRNA or even amino acid sequence is much more complex than it appears. There is therefore a growing mismatch between understanding genes and mRNA (which is growing rapidly) and their resulting proteins (which is growing much more slowly). However, genes and mRNA do not carry out the crucial work required in a cell or organism – proteins do. A key question for bioinformatics is whether this growing gap between knowledge of genes and knowledge of proteins can somehow be bridged by the use of computers.

From this, it can be seen that the process of enzyme/protein production, as determined by DNA, is absolutely critical to the continued well-being of an organism, otherwise organisms as chemical beings would not produce chemical reactions fast enough to keep them alive (e.g. respiration, digestion). Enzymes degrade eventually after catalysing many reactions, and they are broken down into their constituent parts by other enzymes (the 'degradome') for reuse by the organism.

What life now means, according to biomolecular science, is the set of genes (DNA) which code for the production of appropriate enzymes which increase the rate of chemical reactions in cells, where the nature and rate of reactions are determined by the nature of the enzymes. Organisms of a particular species are all essentially the same chemically; what differs are the enzymes produced by the DNA inherited by their parents and other factors (e.g. mutation of individual bases and genes by random means). These enzymes control cellular processes differently for different members of the species, thereby leading to different physical characteristics.

Computers and Biology

Bioinformatics is the symbolic relationship between computational and biological sciences. The ability to sort and extricate genetic codes from a human genomic database of 3 billion base pairs of DNA in a meaningful way is perhaps the simplest form of bioinformatics. Moving on to another level, bioinformatics is useful in mapping different people's genomes and deriving differences in their genetic make-up through pattern recognition software. But that is the easiest part. What is more complex is to decipher the genetic code itself to see what the differences in genetic make-up between different people translate into in terms of physiological traits. And there is yet another level, which is even more intricate and that is the genetic code itself. The genetic code actually codes for amino acids and thereby proteins and the specific role, played by each of these proteins controls the state of our health. The role or function of each of our genes in coding for a specific protein, which in turn regulates a particular metabolic pathway, is described as "functional genomics". The true benefit of bioinformatics therefore lies in harnessing information pertaining to these genetic functions in order to understand how human beings and other living systems operate.

Computational simulation of experimental biology is an important application of bioinformatics, which is referred to as "in silico" testing. This is perhaps an area that will expand in a prolific way, given the need to obtain a greater degree of predictability in animal and human clinical trials. Added to this, is the interesting scope that "in silico" testing provides to deal with the growing hostility towards animal testing. The growth of this sector will largely depend on the acceptance of "in silico" testing by the regulatory authorities. However, irrespective of this, research strategies will certainly find computational modelling to be a vital tool in speeding up research with enormous cost benefits.

Limitations in the Use of Computers

The last decade has witnessed the dawn of a new era of 'silicon-based' biology, opening the door, for the first time, to the possible investigation and comparative analysis of complete genomes. Genome analysis means to elucidate and characterize the genes and gene products of an organism. It depends on a number of pivotal concepts, concerning the processes of evolution (divergence and convergence), the mechanism of protein folding, and the manifestation of protein function.

Today, our use of computers to model such processes is limited by, and must be placed in the context of, the current limits of our understanding of these central themes. At the outset, it is important to recognize that we do not yet fully understand the rules of protein folding; we cannot invariably say that a particular sequence or a fold has arisen by divergent or convergent evolution; and we cannot necessarily diagnose a protein function, given knowledge only of its sequence or of its structure, in isolation. Accepting what we cannot do with computers plays an essential role in forming an appreciation of what, in fact, we can do. Without this kind of understanding, it is easy to be misled, as spurious arguments are often used to promote perhaps rather overenthusiastic points of view about what particular programs and software packages can achieve.

Nature has its own complex rules, which we only poorly understand and which we cannot easily encapsulate within computer programs. No current algorithm can 'do' biology. Programs provide mathematical and therefore infallible, models of biological systems. To interpret correctly whether sequences or structures are meaningfully similar, whether they have arisen by the processes of divergence or convergence, whether similar sequences or similar folds have the same or different functions: these are the most challenging problems. There are no simple solutions, and computers do not give us the answers; rather, given a sea of data, they help to narrow the options down so that the users can begin to draw informed biologically reasonable conclusions.

Current Stage of Research

In the field of bioinformatics, the current research drive is to be able to understand evolutionary relationships in terms of the expression of protein function. Two computational approaches have been brought up to bear on the problem, tackling the identification of protein function from the perspectives of sequence analysis and of structure analysis, respectively. From the point of view of sequence analysis, we are concerned with the detection of relationships between newly determined sequences and those of known function (usually within a database). This may mean pinpointing functional sites shared by disparate proteins (probably the result of convergent evolution), or identifying related functions in similar proteins (most commonly the result of divergent evolution).

The identification of protein function from sequence sounds straight forward, and indeed, sequence analysis is usually a fruitful technique. But, function cannot be inferred from sequence for about one-third of proteins in any of the sequenced genomes, largely because biological characterization cannot keep pace with the volume of data issuing from the genome projects (large number of database sequences thus either carry no annotation beyond the parent gene name, or are simply designated as hypothetical proteins). Another important point is that, in some instances, closely related sequences, which may be assumed to share a common structure, may not share the same function. What this means is that, though sequence or structure analysis can be used for deducing gene functions, still neither technique can be applied infallibly without reference to the underlying biology.

Microbial, Plant and Animal Genomes

Although the human genome appears to be the focal point of interest, microbial, plant and animal genomes are equally exciting to explore through bioinformatics. Mining plant genomics has an important impact on opening up new vistas for research in agriculture. Microbial genomics offers a dual opportunity of developing new fermentation-based products and technologies as well as defining new ways of combating microbial infections. Exploring animal genomics opens up unlimited scope to pursue research in veterinary science and transgenic models.

History (Stages of Development)

The science of sequencing began slowly. The earliest techniques were based on methods for separation of proteins and peptides, coupled with methods for identification and quantification of amino acids. Prior to 1945, there was not a single quantitative analysis available for any one protein. However, significant progress with chromatographic and labelling techniques over the next decade eventually led to the elucidation of the first complete sequence, that of the peptide hormone insulin (1955). Yet, it was the first five years before the sequence of the first enzyme (ribonuclease) was complete (1960). By 1965, around 20 proteins with more than 100 residues had been sequenced, and by 1980, the number was estimated to be of the order of 1500. Today, there are more than 3,00,000 sequences available.

Initially, the manual process of sequential Edman degradation – dansylation, obtained the majority of protein sequences. A key step towards the rapid increase in the number of sequenced proteins was the development of automated sequencers, which, by 1980, offered a 10^4-fold increase in sensitivity relative to the automated procedure implemented by Edman and Begg in 1967.

In the 1960s, scientists struggled to develop methods to sequence nucleic acids. The first technique to emerge was applicable only to tRNA because they are short (74 to 95 nucleotides in length) and it is possible to purify individual molecules.

As against RNA, DNA is very long: human chromosomal molecules may contain between 55 10^6 and 250 10^6 base pairs. Assembling the complete nucleotide sequence of a complete DNA molecule is a huge task. Even if the sequence can be broken down into smaller fragments, purification remains a problem. The advent of gene cloning provided a solution to how the fragments can be separated. By 1977, two sequencing methods had emerged using chain termination and chemical degradation approaches. With only minor changes, the techniques propagated to laboratories throughout the world, and laid the foundation for the sequence revolution of the next two decades, and the subsequent birth of bioinformatics.

During the last decade, molecular biology has witnessed an information revolution as a result of both of the development of rapid DNA sequencing techniques and of the corresponding progress in computer base technologies, which are allowing us to cope with this information deluge in increasingly efficient ways. The broad term that was coined in the mid-1980s to encompass computer applications in biological sciences is bioinformatics. The term Bioinformatics has been commandeered by several different disciplines to mean rather different things. In its broadest sense, the term can be considered to mean information technology applied to the management and analysis of biological sequence data; this has implications in diverse areas, ranging from artificial intelligence and robotics to genome analysis. In the context of genome initiatives, the term was originally applied to the computational manipulation and analysis of biological sequence data. However in view of this recent rapid accumulation of available protein structures, the term now tends also to be used to embrace the manipulation and analysis of 3D structure data.

Applications

Bioinformatics has not only become essential for basic genomic and molecular biology research, but is having a major impact on many areas of biotechnology and biomedical sciences. It has applications, e.g. in knowledge-based drug design, forensic DNA analysis, and agricultural biotechnology. Computational studies of protein–ligand interactions provide a rational basis for the rapid identification of novel leads for synthetic drugs. Knowledge of the three-dimensional structures of proteins allows molecules to be designed that are capable of binding to the receptor site of a target protein with great affinity and specificity. This informatics-based approach significantly reduces the time and cost necessary to develop drugs with higher potency, fewer side effects, and less toxicity than using the traditional trial-and-error approach. In forensics, results from molecular phylogenetic analysis have been accepted as evidence in criminal courts. Some sophisticated Bayesian statistics and likelihood-based methods for analysis of DNA have been applied in the analysis of forensic identity. It is worth mentioning that genomics and bioinformatics are now poised to revolutionize our healthcare system by developing personalized and customized medicine. The high-speed genomic sequencing coupled with sophisticated informatics technology will allow a doctor in a clinic to quickly sequence a patient's genome and easily detect potential harmful mutations and to engage in early diagnosis and effective treatment of diseases. Bioinformatics tools are being used in agriculture as well. Plant genome databases and gene expression profile analyses have played an important role in the development of new crop varieties that have higher productivity and more resistance to disease.

Limitations

Having recognized the power of bioinformatics, it is also important to realize its limitations and avoid over-reliance on and over-expectation of bioinformatics output. In fact, bioinformatics has a number of inherent limitations.

In many ways, the role of bioinformatics in genomics and molecular biology research can be likened to the role of intelligence gathering in battlefields. Intelligence is clearly very important in leading to victory in a battlefield. Fighting a battle without intelligence is inefficient and dangerous. Having superior information and correct intelligence helps to identify the enemy's weaknesses and reveal the enemy's strategy and intentions. The gathered information can then be used in directing the forces to engage the enemy and win the battle. However, completely relying on intelligence can also be dangerous if the intelligence is of limited accuracy. Over reliance on poor-quality intelligence can yield costly mistakes if not complete failures.

It is no stretch in analogy that fighting diseases or other biological problems using bioinformaticsis like fighting battles with intelligence. Bioinformatics and experimental biology are independent, but complementary, activities Bioinformatics depends on experimental science to produce raw data for analysis. It, in turn, provides useful interpretation of experimental data and important leads for further experimental research. Bioinformatics predictions are not formal proofs of any concepts. They do not replace the traditional experimental research methods of actually testing hypotheses. In addition, the quality of bioinformatics predictions depends on the quality of data and the sophistication of the algorithms being used. Sequence data from high throughput analysis often contain errors. If the sequences are wrong or annotations incorrect, the results from the downstream analysis are misleading as well. That is why it is so important to maintain a realistic perspective of the role of bioinformatics.

Genome Projects

In the mid-1980s, the United States Department of Energy initiated a number of projects to construct detailed genetic and physical maps of the human genome, to determine its complete nucleotide sequence, and to localize its estimated 100000 genes. Work on this scale required the development of new computational methods for analysing genetic map and DNA sequence data, and demanded the design of new techniques and instrumentation for detecting and analysing DNA.

To benefit the public most effectively, the projects also necessitated the use of advanced means of information dissemination in order to make the results available as rapidly as possible to scientists and physicians. The international effort arising from this vast initiative became known as the human genome project.

Similar research efforts were also launched to map and sequence the genomes of a variety of organisms used extensively in research labs as model systems. In April 1998, although the sequencing projects of only a small number of relatively small genomes had been completed, and the human genome is not expected to be complete until after the year 2003, the results of such projects were already beginning to pour into the public sequence databases in overwhelming numbers.

Goals for Advancements in Sequencing Technology

DNA sequencing technology has improved dramatically since the genome projects began. The amount of sequence produced each year is increasing steadily; individual centers are now producing tens of millions of base pairs of sequence annually. In the future, de novo sequencing of additional genomes, comparative sequencing of closely related genomes, and sequencing to assess variation within genomes will become increasingly indispensable tools for biological and medical research. Much more efficient sequencing technology will be needed than is currently available. The incremental improvements made to date have not yet resulted in any fundamental paradigm shifts. Nevertheless, the current state-of-the-art technology can still be significantly improved, and resources should be invested to accomplish this. Beyond that, research must be supported on new technologies that will make even higher throughput DNA sequencing efficient, accurate, and cost-effective, thus providing the foundation for other advanced genomic analysis tools. Progress must be achieved in three areas:

a. *Continue to increase the throughput and reduce the cost of current sequencing technology:* Increased automation, miniaturization, and integration of the approaches currently in use, together with incremental, evolutionary improvements in all steps of the sequencing process, are needed to yield further increase in throughput (to at least 500 Mb of finished sequence per year by 2003) and reductions in cost. At least a twofold cost reduction from current levels (which average $0.50 per base for finished sequence in large-scale centers) should be achieved in the next 5 years. Production of the working draft of the human sequence will cost considerably less per base pair.

b. *Support research on novel technologies that can lead to significant improvements in sequencing technology:* New conceptual approaches to DNA sequencing must be supported to attain substantial improvements over the current sequencing paradigm. For example, microelectromechanical systems (MEMS) may allow significant reduction of reagent use, increase in assay speed, and true integration of sequencing functions. Rapid mass spectrometric analysis methods are achieving impressive results in DNA fragment identification and offer the potential for very rapid DNA sequencing. Other more revolutionary approaches, such as single-molecule sequencing methods, must be explored as well. Significant investment in interdisciplinary research in instrumentation, combining chemistry, physics, biology, computer science, and engineering, will be required to meet this goal. Funding of far-sighted projects that may require 5 to 10 years to reach fruition will be essential. Ultimately, technologies that could, for example, sequence one vertebrate genome per year at affordable cost are highly desirable.

c. *Develop effective methods for the advanced development and introduction of new sequencing technologies into the sequencing process:* As the scale of sequencing increases, the introduction of improvements into the production stream becomes more challenging and costly. New technology must therefore be robust and be carefully evaluated and validated in a high-throughput environment before its implementation in a production setting. A strong commitment from both the technology developers and the technology users is essential in this process. It must be recognized that the advanced development process will often require significantly more funds than proof-of-principle studies. Targeted funding allocations and dedicated review mechanisms are needed for advanced technology development.

Developing Technology to Handle Sequence Variations

Natural sequence variation is a fundamental property of all genomes. Any two haploid human genomes show multiple sites and types of polymorphism. Some of these have functional implications, whereas many probably do not. The most common polymo phisms in the human genome are single base-pair differences, also called single-nucleotide polymorphisms (SNPs). When two haploid genomes are compared, SNPs occur every kilobase, on average. Other kinds of sequence variation, such as copy number changes, insertions, deletions, duplications, and rearrangements also exist, but at low frequency and their distribution is poorly understood. Basic information about the types, frequencies, and distribution of polymorphisms in the human genome and in human populations is critical for progress in human genetics. Better high-throughput methods for using such information in the study of human disease are also needed.

SNPs are abundant, stable, widely distributed across the genome, and lend themselves to automated analysis on a very large scale, e.g. with DNA array technologies. Because of these properties, SNPs will be a boon for mapping complex traits such as cancer, diabetes, and mental illness. Dense maps of SNPs will make possible genome-wide association studies, which are a powerful method for identifying genes that make a small contribution to disease risk.

In some instances, such maps will also permit prediction of individual differences in drug response. Publicly available maps of large numbers of SNPs distributed across the whole genome, together with technology for rapid, large-scale identification and scoring of SNPs, must be developed to facilitate this research.

a. *Develop technologies for rapid, large-scale identification of SNPs and other DNA sequence variants:* The study of sequence variation requires efficient technologies that can be used on a large scale and that can accomplish one or more of the following tasks: rapid identification of many thousands of new SNPs in large numbers of samples. Although the immediate emphasis is on SNPs, ultimately technologies that can be applied to polymorphisms of any type must be developed. Technologies are also needed that can rapidly compare, by large-scale identification of similarities and differences, the DNA of a species that is closely related to one whose DNA has already been sequenced. The technologies that are developed should be cost-effective and broadly accessible.
b. *Identify common variants in the coding regions of the majority of identified genes:* Initially, association studies involving complex diseases will likely test a large series of candidate genes; eventually, sequences in all genes may be systematically tested. SNPs in coding sequences (also known as cSNPs) and the associated regulatory regions will be immediately useful as specific markers for disease. An effort should be made to identify such SNPs as soon as possible. Ultimately, a catalog of all common variants in all genes will be desirable. This should be cross-referenced with cDNA sequence data.
c. *Create an SNP map of at least 100,000 markers:* A publicly available SNP map of sufficient density and informativeness to allow effective mapping in any population is the ultimate goal. A map of 100,000 SNPs (one SNP per 30,000 nucleotides) is likely to be sufficient for studies in some relatively homogeneous populations, while denser maps may be required for studies in large, heterogeneous populations.

Thus, during this 5-year period, the HGP authorities have planned to create a map of at least 100,000 SNPs. If technological advances permit, a map of greater density is desirable. Research should be initiated to estimate the number of SNPs needed in different populations.

d. *Develop the intellectual foundations for studies of sequence variation:* The methods and concepts developed for the study of single-gene disorders are not sufficient for the study of complex, multigene traits. The study of the relationship between human DNA sequence variation, phenotypic variation, and complex diseases depends critically on better methods. Effective research design and analysis of linkage, linkage disequilibrium, and association data are areas that need new insights. Questions such as which study designs are appropriate to which specific populations, and with which population genetics characteristics, must be answered. Appropriate statistical and computational tools and rigorous criteria for establishing and confirming associations must also be developed.

e. *Create public resources of DNA samples and cell lines:* To facilitate SNP discovery, it is critical that common public resources of DNA samples and cell lines be made available as rapidly as possible. To maximize discovery of common variants in all human populations, a resource is needed that includes individuals whose ancestors derive from diverse geographic areas. It should encompass as much of the diversity found in the human population as possible. Samples in this initial public repository should be totally anonymous to avoid concerns that arise with linked or identifiable samples.

DNA samples linked to phenotypic data and identified as to their geographic and other origins will be needed to allow studies of the frequency and distribution of DNA polymorphisms in specific populations and their relevance to disease. However, such collections raise many ethical, legal, and social concerns that must be addressed. Credible scientific strategies must be developed before creating these resources.

Need of Technology in Functional Genomics

Functional genomics is the interpretation of the function of DNA sequence on a genomic scale. Already, the availability of the sequence of entire organisms has demonstrated that many genes and other functional elements of the genome are discovered only when the full DNA sequence is known. Such discoveries will accelerate as sequence data accumulate. However, knowing the structure of a gene or other element is only part of the answer. The next step is to elucidate function, which results from the interaction of genomes with their environment. Current methods for studying DNA function on a genomic scale include comparison and analysis of sequence patterns directly to infer function, large-scale analysis of the messenger RNA and protein products of genes, and various approaches to gene disruption. In the future, a host of novel strategies will be needed for elucidating genomic function. This will be a challenge for all of biology. The HGP will be contributing to this area by emphasizing the development of technology that can be used on a large scale, is efficient, and is capable of generating complete data for the genome as a whole. To the extent that available resources allow, expansion of current approaches as well as innovative technology ideas should be supported in the areas described below.

a. *Develop cDNA resources:* Complete sets of full-length cDNA clones and sequences for both humans and model organisms would be enormously useful for biologists and are urgently needed. Such resources would help in both gene discovery and functional analysis. High priority should be placed on developing technology for obtaining full-length cDNAs. Complete and validated inventories of full-length cDNA clones and corresponding sequences should be generated and made available to the community once such technology is at hand.

b. Improved technologies are needed for global approaches to the study of non-protein-coding sequences, including production of relevant libraries, comparative sequencing, and computational analysis.

c. Develop technology for comprehensive analysis of gene expression. Information about the spatial and temporal patterns of gene expression in both humans and model organisms offers one key to understanding gene expression.

   Efficient and cost-effective technology needs to be developed to measure various parameters of gene expression reliably and reproducibly. Complementary DNA sequences and validated sets of clones with unique identifiers will be needed for array technologies, large-scale in situ hybridization, and other strategies for measuring gene expression. Improved methods for quantifying, representing, analyzing, and archiving expression data should also be developed.

d. Improve methods for genome-wide mutagenesis. Creating mutations that cause loss or alteration of function is another prime approach to studying gene function. Technologies, both gene- and phenotype-based, which can be used on a large scale in vivo or in vitro, are needed for generating or finding such mutations in all genes. Such technologies should be piloted in appropriate model systems, including both cell culture and whole organisms.

e. Develop technology for global protein analysis. A full understanding of genome function requires an understanding of protein function on a genome-wide basis. Development of experimental and computational methods to study global spatial and temporal patterns of protein expression, protein-ligand interactions, and protein modification needs to be supported.

2

Biological Databases

What is a Database?

A database is a computerized archive used to store and organize data in such a way that information can be retrieved easily via a variety of search criteria. Databases are composed of computer hardware and software for data management. The chief objective of the development of a database is to organize data in a set of structured records to enable easy retrieval of information. Each record, also called an entry, should contain a number of fields that hold the actual data items, e.g. fields for names, phone numbers, addresses, dates. To retrieve a particular record from the database, a user can specify a particular piece of information, called value, to be found in a particular field and expect the computer to retrieve the whole data record. This process is called making a query.

Although data retrieval is the main purpose of all databases, biological databases often have a higher level of requirement, known as knowledge discovery, which refers to the identification of connections between pieces of information that were not known when the information was first entered. For example, databases containing raw sequence information can perform extra-computational tasks to identify sequence homology or conserved motifs. These features facilitate the discovery of new biological insights from raw data.

Biological Sequence/Structure Deficit

At the beginning of 1998, in publicly available, non-redundant databases, more than 3,00,000 protein sequences have been deposited, and the number of partial sequences in public and proprietary expressed sequence tag databases is estimated to run into millions. By contrast, the number of unique 3D structures in the protein data bank (PDB) was less than 1500. Although structural information is far more complex to derive, store and manipulate than are sequence data, these figures nevertheless highlight an enormous information deficit. This situation is likely to get worse as the genome projects around the world begin to bear fruit. Of course, the acquisition of structural data is also hastening, and the future large-scale structure determination enterprise could conceivably furnish 2000 3D structures annually. But this is a small yield by comparison with that of sequence databases, which are doubling in size every year, with a new sequence being added, on average once a minute.

Biological Databases

If we are to derive the maximum benefit from the deluge of sequence information, we must deal with it in a concerted way; this means establishing, maintaining and disseminating databases; providing easy to use software to access the information they contain; and designing state-of-the-art analysis tools to visualize and interpret the structural and functional clues latent in the data.

The first, then, in analysing sequence information is to assemble it into central shareable resources, i.e. databases. Databases are effectively electronic filling cabinets, a convenient and efficient method of storing vast amounts of information. There are many different database types, depending both on the nature of the information being stored and on the manner of data storage (e.g. whether in flat-files, tables in a relational database or objects in an object-oriented database).

In the context of protein sequence analysis, we will encounter primary, composite and secondary databases. Such resources store different levels of information in totally different formats.

In the past, this has led to a variety of communication problems, but emerging computer technologies are beginning to provide solutions, allowing seamless, transparent access to disparate, distributed data structures over the internet.

Primary and secondary databases are used to address different aspects of sequence analysis, because they store different levels of protein sequence information.

The primary structure of a protein is its amino acid sequence; these are stored in primary databases as linear alphabets that denote the constituent residues. The secondary structure of a protein corresponds to regions of local regularity, which, in sequence alignments, are often apparent as well conserved motifs; these are stored in secondary databases as patterns. The tertiary structure of a protein arises from the packing of its secondary structure elements which may form discrete domains within a fold, or may give rise to autonomous folding units or modules; complete folds, domains and modules are stored in structure databases as sets of atomic co-ordinates.

Primary Sequence Databases

In the early 1980s, sequence information started to become more abundant in the scientific literature. Realising this, several laboratories saw that there might be advantages to harvesting and storing these sequences in central repositories. Thus, several primary database projects began to evolve in different parts of the world.

Nucleic Acid Sequence Databases

The principle DNA sequence databases are GenBank (USA), EMBL (Europe) and DDBJ (Japan), which exchange data on a daily basis to ensure comprehensive coverage at each of the sites.

GenBank

GenBank, the DNA database from the National Center for Biotechnology Information, exchanges data with both EMBL and DDBJ to help ensure comprehensive coverage. The database is split into 17 smaller discrete divisions.

The GenBank flatfile (GBFF) is the elementary unit of information in the GenBank database. It is one of the most commonly used formats in the representation of biological sequences.

The GBFF can be separated into three parts: the header, which contains the information (descriptors) that apply to the whole record; the features, which are the annotations on the record; and the nucleotide sequence itself. All major nucleotide database flat files end with // on the last line of the record.

```
LOCUS       CAA00094                 54 aa            linear   PAT 14-NOV-2006
DEFINITION  amylase [Aspergillus niger].
ACCESSION   CAA00094
VERSION     CAA00094.1  GI:14646
DBSOURCE    embl accession A00845.1
KEYWORDS    .
SOURCE      Aspergillus niger
  ORGANISM  Aspergillus niger
            Eukaryota; Fungi; Dikarya; Ascomycota; Pezizomycotina;
            Eurotiomycetes; Eurotiomycetidae; Eurotiales; Trichocomaceae;
            mitosporic Trichocomaceae; Aspergillus.
REFERENCE   1  (residues 1 to 54)
  AUTHORS   Woeldike,H.
  TITLE     PROCESS FOR THE PRODUCTION OF PROTEIN PRODUCTS IN ASPERGILLUS AND
            PROMOTERS FOR USE IN ASPERGILLUS
  JOURNAL   Patent: WO 8901969-A 09-MAR-1989;
            Novo Industries
FEATURES             Location/Qualifiers
     source          1..54
                     /organism="Aspergillus niger"
                     /db_xref="taxon:5061"
     Protein         1..54
                     /product="amylase"
     mat_peptide     17..54
                     /product="amylase"
     CDS             1..54
                     /coded_by="A00845.1:1..162"
ORIGIN
        1 mtiflflaif vatalaatpa ewrsqsiyfl ltdrfartdn sttascdlsa rvsh
//
```

Accession Number or Flatfile Number

Date of Submission

Original Sequence

Fig. 2.1: GenBank flatfile format

The Header

The header is the most database-specific part of the record. The first line of all GBFFs is the locus line.

Locus

It consists of accession number for the sequence, length of the sequence, molecule type, GenBank division code and date the record was last made public

Definition

The definition line (also referred to as the "def line") is the line in the GenBank record that attempts to summarize the biology of the record. This is the line that appears in the FASTA files that NCBI generates and is what is seen in the summary line for BLAST hits generated from a BLAST similarity search.

Accession

The accession number, on the third line of the record, represents the primary key to reference a given record in the database. This is the number that is cited in publications and is always associated with this record; that is, if the sequence is updated (e.g. by changing a single nucleotide), the accession number will not change. All GenBank records have only a single line with the word ACCESSION on it; however, there may be more than one accession number. The vast majority of records only have one accession number. This number is always referred to as the primary accession number; all others are secondary. In cases where more than one accession number is shown, the first accession number is the primary one.

Key Word

Adding key words to an entry is often not very useful because over the years so many authors have selected words not on a list of controlled vocabulary and not uniformly applied to the whole database. NCBI, therefore, discourages the use of key words but will include them on request, especially if the words are not present elsewhere in the record or are used in a controlled fashion (e.g. for EST, STS, GSS, and HTG records).

Source

The source line will either have the common name for the organism or its scientific name.

Reference

Each GenBank record must have at least one reference or citation. It offers scientific credit and sets a context explaining why this particular sequence was determined.

The Feature Table

The middle segment of the GBFF record, the feature table, is the most important direct representation of the biological information in the record. In the NCBI data model, "features" refer to annotations that are on a part of the sequences, whereas annotations that describe the whole sequence are called "descriptors." Thus, the source feature in the GenBank flatfile is really a descriptor in the data model view (the BioSource, which refers to the whole sequence), not a feature as used elsewhere.

The Source Feature

The source feature is the only feature that must be present on all GenBank records. All features have a series of legal qualifiers, some of which are mandatory (e.g. /organism for source). All DNA sequence records have some origin, even if synthetic in the extreme case. In most cases, there will be a single source feature, and it will contain the organism. The organism qualifier contains the scientific genus and species name. In some cases, "organisms" can be described at the subspecies level. For the source feature, the series of qualifiers will contain all matters relating to the BioSource, and these may include mapping, chromosome or tissue from which the molecule that was sequenced was obtained, clone identification, and other library information.

Implicit in the BioSource and the organism that is assigned to it is the genetic code used by the DNA/RNA, which will be used to translate the nucleic acid to represent the protein sequence (if one is present in the record). This information is shown on the CDS feature. The CDS feature contains instructions to the reader on how to join two sequences together or on how to make an amino acid sequence from the indicated coordinates and the inferred genetic code.

Three letter codes for each of divisions in GenBank.

- PRI : Primate
- ROD: Rodent
- MAM: Mammalian
- VRT: Vertebrate
- INV: Invertebrate
- PLN: Plant, fungal, algal
- BCT: Bacterial
- RNA: Structural RNA
- VRL: Viral
- HTG: High throughput genome sequences
- PHG: Bacteriophage
- SYN: Synthetic
- UNA: Unannotated
- EST: Expressed sequence tag
- PAT: Patent
- STS: Sequence tagged sites
- GSS: Genome survey sequences

EMBL

The European Molecular Biology Laboratory (EMBL) maintains DNA and protein sequence databases. The rate of growth of DNA databases has been following an exponential trend, with a doubling time less than a year. EMBL data predominantly (more than 50%) consist of model organisms. The format for each entry in these databases is shown in Fig. 2.2.

As with GenBank entries, a large amount of information describing each sequence entry is given, including literature references, information about the function of the sequence, locations of mRNAs and coding regions, and positions of important mutations. This information is organized into fields, each with an identifier, shown as the first text on each line. These identifiers are abbreviated to two letters, e.g. RF for reference, and some identifiers may have additional subfields. The sequence entry is assumed by computer programs to lie between the identifiers "SEQUENCE" and "//" and includes numbers on each line to locate parts of the sequence visually.

DNA Data Bank of Japan (DDBJ)

DNA Data Bank of Japan is produced, distributed and maintained by the National Institute of Genetics.

Protein Sequence Databases

PIR, MIPS, SWISS-PROT, and TrEMBL are the major protein sequence databases.

PIR was developed for investigating evolutionary relations between proteins. In its current form, the database is split into four distinct sections PIR1–PIR4, which differ in terms of the quality of data and the level of annotation provided.

```
ID        identifications code for sequence in the database
AC        accession number giveing origin of sequence
DT        dates of entry and modification
KW        key cross-reference words for lookup up this entry
OS, OC    source organism
RN, RP, RX, RA, RT, RL literature reference or source
DR        I.d. in other databses
CC        description of biological function
FH, FT    infromatin about sequence by base position or range of positions
          source range of sequence, source organism
          misc_signal range of sequence, type of function or signal
          mRNA range of sequence, mRNA
          CDS range of sequence, protein coding region
          intron range of sequence, position of intron
          mutatin sequence position, change in sequence for mutation
SQ        count of A, C, G, T and other symbols
gaattogata aatctctggt ttattgtgca gtttatggtt ccaaaatcgo ottttgctgt 60
atatactacac agcataactg tatatacacc catgggggcgg aatgaaagcg ttaacggcca 120
:
// symbol ot indicate end of sequence
```

Fig. 2.2: EMBL file format

This sequence format, which is sometimes also called the PIR format, has been used by the National Biomedical Research Foundation/ Protein Information Resource (NBRF) and also by other sequence analysis programs. The NBRF format is similar to the FASTA sequence format but with significant differences. The first line includes an initial "_" character followed by a two-letter code such as P for complete sequence or F for fragment, followed by a 1 or 2 to indicate type of sequence, then a semicolon, then a four- to six-character unique name for the entry. There is also an essential second line with the full name of the sequence, a hyphen, then the species of origin.

In FASTA format, the second line is the start of the sequence and the first line gives the sequence identifier after a "_" sign. The sequence terminates with an asterisk.

```
>P1;ILEC
lexA repressor - Escherichia coli
MKALTARQQEVFDLIRDHISQTGMPPTRAE
IAQRLGFRSPNAAEEHLKALARKGVIEIVS
GASRGIRLLQEEEEGLPLVGRVAAGEPLLA
QQHIEGHYQVDPSLFKPNADFLLRVSGMSM
KDIGIMDGDLLAVHKTQDVRNGQVVVARID
DEVTVKRLKKQGNKVELLPENSEFKPIVVD
LRQQSFTIEGLAVGVIRNGDWL
```

Fig. 2.3: NBRF sequence entry format

MIPS collects and processes sequence data for the tripartite PIR-International Protein sequence Database Project.

SWISS-PROT is a protein sequence database which, endeavors to provide high level annotations, including descriptions of the function of the protein, and of the structure of its domain, its post-translational modifications, and so on. The format of an entry in the SWISS-PROT protein sequence database is very similar to the EMBL format, except that considerably more information about the physical and biochemical properties of the protein is provided.

TrEMBL was created as a supplement to the SWISS-PROT. It was designed to address the need for a well-structured SWISS-PROT-like resource that would allow very rapid access to sequence data from the genome projects, without having to compromise the quality of SWISS-PROT itself by incorporating sequences with insufficient analysis and annotation.

Composite Protein Sequence Databases

One solution to the problem of proliferation primary databases is to compile a composite, i.e. a database that amalgamates a variety of different primary sources. Composite databases render sequence searching much more efficient, because they obviate the need to interrogate multiple resources. The interrogation process is stream-lined still further if the composite has been designed to be non redundant, as this means that the same sequence need not be searched more than once. The choices of different sources and the application of different redundancy criteria have led to the emergence of different composites. The major composite databases are non-redundant database, OWL, MIPSX, SWISS-PROT+TrMBL.

Secondary Databases

Secondary databases contain the fruits of analyses of the sequences in the primary resources. Because there are several different primary databases and a variety of ways of analysing protein sequences, the information housed in each of the secondary resources is different. Designing software tools that can search the different types of data, interpret the range of outputs, and assess the biological significance of the results is not a trivial task. SWISS-PROT has emerged as the most popular primary source and many secondary databases now use it as their basis.

Some of the main secondary resources are mentioned in Table 2.1.

Table 2.1: Secondary structure resources

Secondary database	*Primary source*	*Stored information*
Prosite	SWISS-PROT	Regular expressions
Profiles	SWISS-PROT	Weighted matrices
Prints	OWL	Aligned motifs
Pfam	SWISS-PROT	Hidden Marcov Models
Blocks	PROSITE/PRINTS	Aligned motifs (blocks)
Identify	BLOCKS/PRINTS	Fuzzy regular expressions

Tertiary Databases

Tertiary databases are the databases derived from information housed in secondary (pattern) databases (e.g. the BLOCKS and eMOTIF databases, which draw on the data stored within PROSITE and PRINTS). The value of such resources is in providing a different scoring perspective on the same underlying data, allowing the possibility to diagnose relationships that might be missed using the original implementation.

Structural Classification of Proteins

The structural classification of proteins (SCOP) database is a largely manual classification of protein structural domains based on similarities of their amino acid sequences and three-dimensional structures. Originally published in 1995, it is usually updated at least once yearly by Alexei G. Murzin and his colleagues, upon whose expertise the classification rests.

SCOP utilizes four levels of hierarchic structural classification:

1. Class—general "structural architecture" of the domain.
2. Fold—similar arrangement of regular secondary structures but without evidence of evolutionary relatedness.
3. Superfamily—sufficient structural and functional similarity to infer a divergent evolutionary relationship but not necessarily detectable sequence homology.
4. Family—some sequence similarity can be detected.

Comparison to other classification systems

This classification is more significantly based on the human expertise than semi-automatic CATH, its chief rival. It is usually accepted that SCOP provides a better justified classification. Human expertise is needed to decide whether certain proteins are evolutionary related and therefore should be assigned to the same superfamily, or their similarity is a result of structural constraints and therefore they belong to the same fold. Another database, FSSP, is purely automatically generated (including regular automatic updates) but offers no classification, allowing the user to draw their own conclusion as to the significance of structural relationships based on the pairwise comparisons of individual protein structures.

Structural classes

SCOP includes the following structural classes:

1. α-helical domains
2. β-sheet domains
3. α/β domains which consist of from "beta-alpha-beta" structural units or "motifs" that form mainly parallel β-sheets
4. α+β domains formed by independent α-helices and mainly antiparallel β-sheets
5. Multi-domain proteins
6. Membrane and cell surface proteins and peptides (not including those involved in the immune system)
7. "Small" proteins
8. Coiled-coil proteins
9. Low-resolution protein structures
10. Peptides and fragments
11. Designed proteins of non-natural sequence.

Medical Databases

PubMed

PubMed is a free search engine for accessing the MEDLINE database of citations, abstracts and some full text articles on life sciences and biomedical topics. The United States National Library of Medicine at the National Institutes of Health maintains PubMed as part of the Entrez information retrieval system. Listing an article or journal in PubMed is not endorsement. In addition to MEDLINE, PubMed also offers access to

- OLDMEDLINE for pre-1966 citations: This has recently been enhanced, and records for 1951+, even those parts in the printed indexes, are now included within the main portion.

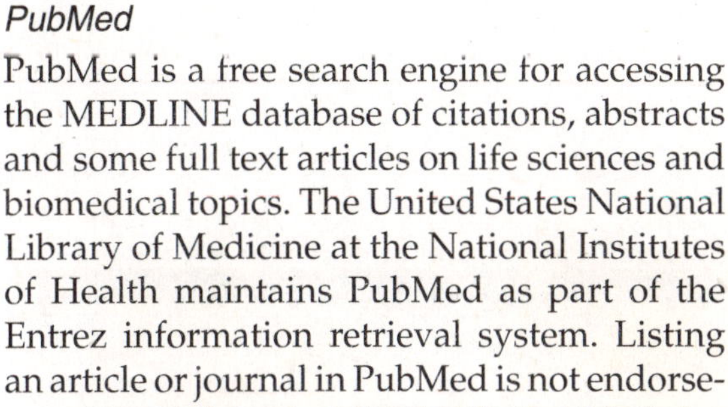

- Citations to all articles, even those that are out-of-scope (e.g. covering plate tectonics or astrophysics) from certain MEDLINE journals, primarily the most important general science and chemistry journals, from which the life science articles are indexed for MEDLINE.
- In-process, citations which provide a record for an article before it is indexed with MeSH and added to MEDLINE or converted to out-of-scope status.
- Citations that precede the date that a journal was selected for MEDLINE indexing (when supplied electronically by the publisher).
- Some life science journals that submit full text to PubMed Central and may not have been recommended for inclusion in MEDLINE although they have undergone a review by NLM, and some physics journals that were part of a prototype PubMed in the early to mid-1990s.

Many PubMed citations contain links to full text articles which are freely available, often in the PubMed Central Digital Library. In late 2007, President George W. Bush signed the Consolidated Appropriations Act of 2007 into law; this law included a provision requiring the NIH to modify its policies and require inclusion into PubMed Central complete electronic copies of their peer-reviewed research and findings from its funded research. This is the first time the US government has required an agency to provide open access to research and is an evolution from the 2005 policy, in which the NIH asked researchers to voluntarily add their research to PubMed Central. With an effective date of April 7, 2008, the Department of Health and Human Services has given notice: "The Director of the National Institutes of Health shall require that all investigators funded by the NIH submit or have submitted for them to the National Library of Medicine's PubMed Central an electronic version of their final, peer-reviewed manuscripts upon acceptance for publication, to be made publicly available no later than 12 months after the official date of publication: Provided, that the NIH shall implement the public access policy in a manner consistent with copyright law."

PubMed is one of a number of search engines through which it is possible to search the MEDLINE database; the National Library of Medicine also leases the MEDLINE information to a number of private vendors such as Ovid and SilverPlatter – as well as many other vendors. PubMed has been available free on the Internet since the mid-1990s.

Information about the journals indexed in PubMed is found in its journals database, searchable by subject or journal title, title abbreviation, the NLM ID (NLM's unique journal identifier), the ISO abbreviation, and both the print and electronic International Standard Serial Numbers (pISSN and eISSN). The database includes all journals in all Entrez databases.

OMIM

Online Mendelian Inheritance in Man (OMIM) project is a database that catalogues all the known diseases with a genetic component, and when possible links them to the relevant genes in the human genome and provides references for further research and tools for genomic analysis of a catalogued gene.

Versions

It is available as a book named after the project, and it is currently in its 12th edition. The online version is called Online Mendelian Inheritance in Man (OMIM), which can

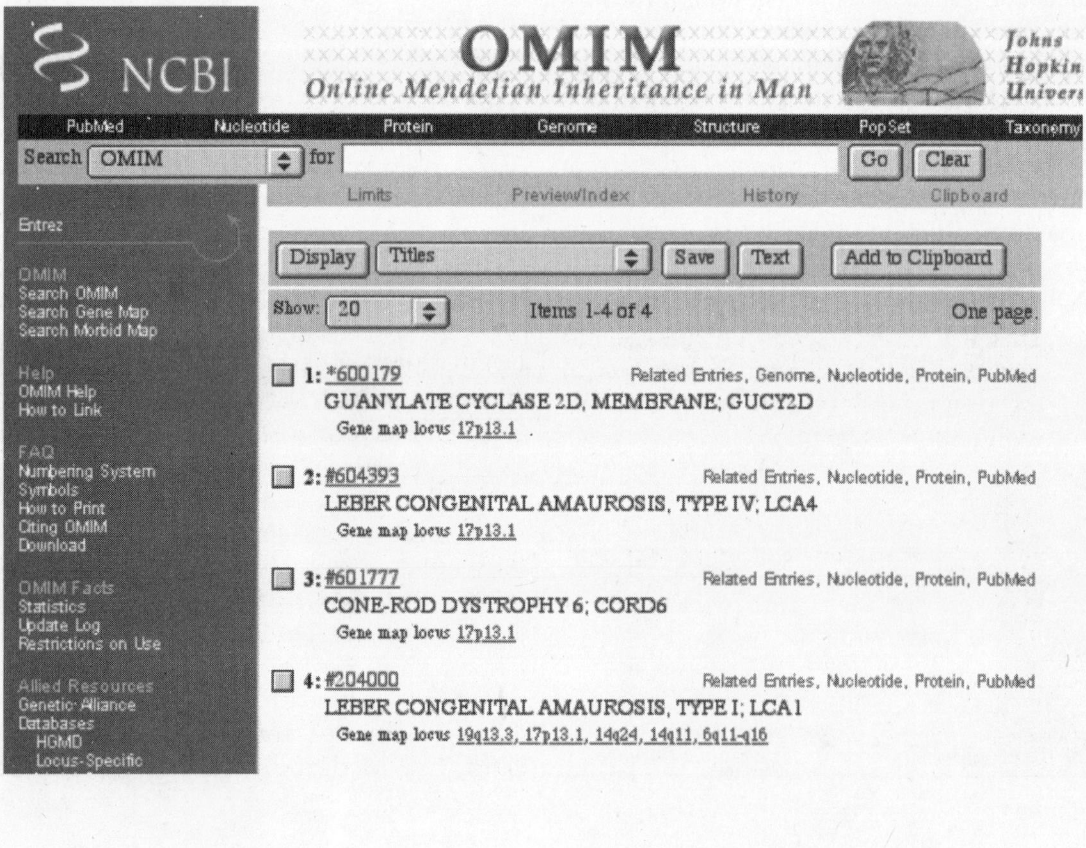

be accessed with the Entrez database searcher of the National Library of Medicine and is part of the NCBI Education project.

Collection process

The information in this database was collected and processed under the leadership of Dr. Victor A. McKusick at Johns Hopkins University, assisted by a team of science writers and editors. Relevant articles are identified, discussed and written up in the relevant entries in the MIM database.

The MIM code

Every disease and gene is assigned a six-digit number of which the first number classifies the method of inheritance.

If the initial digit is 1, the trait is deemed autosomal dominant; if 2, autosomal rece sive; if 3, X-linked. Wherever a trait defined in this dictionary has a MIM number, the number from the 12th edition of MIM, is given in square brackets with or without an asterisk (asterisks indicate that the mode of inheritance is known; a number symbol (#) before an entry number means that the phenotype can be caused by mutation in any of two or more genes) as appropriate, e.g. Pelizaeus–Merzbacher disease [MIM* 169500] is a well-established, autosomal, dominant, mendelian disorder.

Structure Databases

Protein Data Bank (PDB)

The protein data bank (PDB) is a repository for the 3-D structural data of large biological molecules, such as proteins and nucleic acids. The data, typically obtained by X-ray crystallography or NMR spectroscopy and submitted by biologists and biochemists from around the world, can be accessed at no charge on the internet. The PDB is overseen by an organization called the worldwide protein Data Bank, wwPDB.

The PDB is a key resource in areas of structural biology, such as structural genomics. Most major scientific journals, and some funding agencies, such as the NIH in the USA, now require scientists to submit their structure data to the PDB. If the contents of the PDB are thought of as primary data, then there are hundreds of derived (i.e. secondary) databases that categorize the data differently. For example, both SCOP and CATH categorize structures according to type of structure and assumed evolutionary relations; GO categorize structures based on genes.

History

The PDB originated as a grassroots effort. In 1971, Walter Hamilton of the Brookhaven National Laboratory agreed to set up the data

Table 2.2: Representation of OMIM code

First digit	*Range of MIM codes*	*Method of inheritance*
1	100000–199999	Autosomal dominant loci or phenotypes (created before May 15, 1994)
2	200000–299999	Autosomal recessive loci or phenotypes (created before May 15, 1994)
3	300000–399999	X-linked loci or phenotypes
4	400000–499999	Y-linked loci or phenotypes
5	500000–599999	Mitochondrial loci or phenotypes
6	600000–	Autosomal loci or phenotypes (created after May 15, 1994)

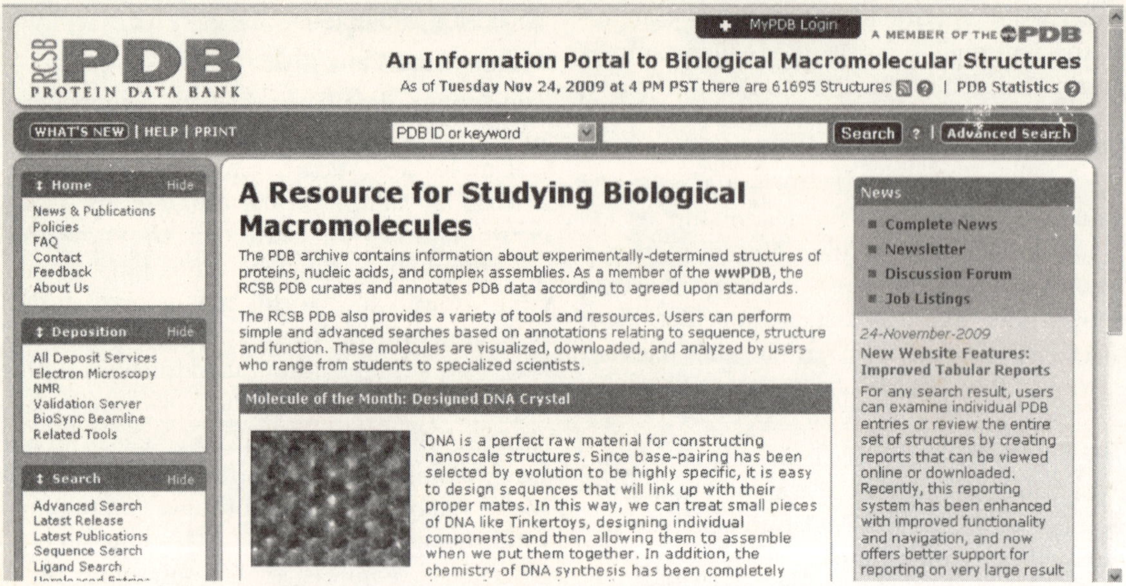

bank at Brookhaven. Upon Hamilton's death in 1973, Tom Koeztle took over direction of the PDB. In January, 1994, Joel Sussman was appointed head of the PDB. In October, 1998, the PDB was transferred to the Research Collaboratory for Structural Bioinformatics (RCSB); the transfer was completed in June, 1999. The new director was Helen M. Berman of Rutgers University (one of the member institutions of the RCSB). In 2003, with the formation of the wwPDB, the PDB became an international organization. Each of the four members of wwPDB can act as deposition, data processing and distribution centers for PDB data. The data processing refers to the fact that wwPDB staff review and annotates the each submitted entry. The data are then automatically checked for plausibility. (The source code for this validation software has been made available to the public at no charge.)

File Format

The file format initially used by the PDB was called the PDB file format. This original format was restricted by the width of computer punch cards to 80 characters per line. Around 1996, the "macromolecular Crystallographic Information file" format, mmCIF, started to be phased in. An XML version of this format, called PDBML, was described in 2005. The structure files can be downloaded in any of these three formats. In fact, individual files are easily downloaded into graphics packages using web addresses:

- For PDB format files, use, e.g. http://www.pdb.org/pdb/files/4hhb.pdb.gz
- For PDBML (XML) files, use, e.g. http://www.pdb.org/pdb/files/4hhb.xml.gz

The "4hhb" is the PDB identifier. Each structure published in PDB receives a four-character alphanumeric identifier, its PDB ID. (This cannot be used as an identifier for biomolecules, because often several structures for the same molecule (in different environments or conformations) are contained in PDB with different PDB IDs.)

Viewing the Data

The structure files may be viewed using one of several open source computer programs. Some other free, but not open source programs include VMD, MDL Chime, Swiss-PDB Viewer, StarBiochem (a Java-based interactive molecular viewer with integrated search of Protein Data Bank) and Sirius. The RCSB PDB website contains an extensive list of both free and commercial molecule visualization programs and web browser plugins.

Protein Data Bank (file format)

The Protein Data Bank (pdb) file format is a textual file format describing the three-dimensional structures of molecules held in the Protein Data Bank. Most of the information in that database pertains to proteins, and the pdb format accordingly provides for rich description and annotation of protein properties. However, proteins are often crystallized in association with other molecules or ions such as water, ions, nucleic acids, drug molecules and so on, which therefore can be described in the pdb format as well.

A typical pdb file describing a protein consists of hundreds to thousands of lines like the following (taken from a file describing the structure of a synthetic collagen-like peptide.

The ATOM records describe the coordinates of the atoms that are part of the protein. For example, the first ATOM line above describes the alpha-N atom of the first residue of peptide chain A, which is a proline residue; the first three floating point numbers are its x, y and z coordinates. What are the next two numbers? HETATM records describe coordinates of hetero-atoms, that is those atoms which are not part of the protein molecule. The SEQRES records give the sequences of the three peptide chains (named A, B and C), which are very short in this example but usually span multiple lines. REMARK records can contain free-form annotation, but they also accommodate standardized information; for example, the REMARK

```
HEADER    EXTRACELLULAR MATRIX                                  22-JAN-98   1A3I
TITLE     X-RAY CRYSTALLOGRAPHIC DETERMINATION OF A COLLAGEN-LIKE
TITLE     2 PEPTIDE WITH THE REPEATING SEQUENCE (PRO-PRO-GLY)
...
EXPDTA    X-RAY DIFFRACTION
AUTHOR    R.Z.KRAMER,L.VITAGLIANO,J.BELLA,R.BERISIO,L.MAZZARELLA,
AUTHOR    2 B.BRODSKY,A.ZAGARI,H.M.BERMAN
...
REMARK 350 BIOMOLECULE: 1
REMARK 350 APPLY THE FOLLOWING TO CHAINS: A, B, C
REMARK 350     BIOMT1  1   1.000000   0.000000    0.000000    0.00000
REMARK 350     BIOMT2  1   0.000000   1.000000    0.000000    0.00000
...
SEQRES    1    A    9   PRO PRO GLY PRO PRO GLY PRO PRO GLY
SEQRES    1    B    6   PRO PRO GLY PRO PRO GLY
SEQRES    1    C    6   PRO PRO GLY PRO PRO GLY
...
ATOM      1    N    PRO A   1    8.316  21.206  21.530 1.00 17.44 N
ATOM      2    CA   PRO A   1    7.608 20.729  20.336 1.00 17.44 C
ATOM      3    C    PRO A   1    8.487  20.707  19.092 1.00 17.44 C
ATOM      4    O    PRO A   1    9.466 21.457  19.005  1.00 17.44 O
ATOM      5    CB   PRO A   1    6.460  21.723  20.211  1.00 22.26 C
...
HETATM  130    C    ACY    401   3.682 22.541  11.236  1.00 21.19 C
HETATM  131    O    ACY    401   2.807 23.097  10.553  1.00 21.19 O
HETATM  132    OXT  ACY    401   4.306 23.101  12.291  1.00 21.19 O
...
```

Fig. 2.3: PDB flat file format

350 BIOMT records describe how to compute the coordinates of the experimentally observed multimer from those of the explicitly specified ones of a single repeating unit.

HEADER, TITLE and AUTHOR records provide information about the researchers who defined the structure; numerous other types of records are available to provide other types of information.

Other Databases

EST Database

Expressed sequence tags (ESTs) are short (usually about 300–500 bp), single-pass sequence reads from mRNA (cDNA). Typically they are produced in large batches. They represent a snapshot of genes expressed in a given tissue and/or at a given developmental stage. They are tags (some coding, others not) of expression for a given cDNA library.

dbEST is a division of GenBank that contains sequence data and other information on "single-pass" cDNA sequences, or "Expressed sequence tags", from a number of organisms.

Access to dbEST Data

- EST sequences are included in the EST division of GenBank, available from NCBI by anonymous ftp and through Entrez.
- The nucleotide sequences may be searched using the BLAST electronic mail server. The TBLASTN program which takes an amino acid query sequence and compares it with six-frame translations of dbEST DNA sequences is particularly useful.
- EST sequences are also available as a flat file in the FASTA format by anonymous ftp in the /repository/dbEST directory at ftp.ncbi.nih.gov

http://www.ncbi.nlm.nih.gov/dbEST/dbEST_access.html

Most EST projects develop large numbers of sequences. These are commonly submitted to GenBank and dbEST as batches of dozens to thousands of entries, with a great deal of redundancy in the citation, submitter and library information. To improve the efficiency of the submission process for this type of data, a special streamlined submission process and data format has been designed.

dbEST also includes sequences that are longer than the traditional ESTs, or are produced as single sequences or in small batches. Among these, sequences are products of differential display experiments and RACE experiments. The thing that these sequences have in common with traditional ESTs, regardless of length, quality, or quantity, is that there is little information that can be annotated in the record.

If a sequence is later characterized and annotated with biological features such as a coding region, 5'UTR, or 3'UTR, it should be submitted through the regular GenBank submissions procedure (via BankIt or Sequin), even if part of the sequence is already in dbEST.

dbEST is reserved for single-pass reads. Assembled sequences should not be submitted to dbEST. GenBank will accept assembled EST submissions for the forthcoming TSA (Transcriptome Shotgun Assembly) division. More information about submitting EST assemblies can be got by writing to gb-admin@ncbi.nlm.nih.gov.

The individual reads which make up the assembly should be submitted to dbEST, the Trace archive or the short read archive (SRA) prior to the submission of the assemblies.

Sequences which should not be included in EST submissions include the following: mitochondrial sequences, rRNA, viral sequences, vector sequences. Vector and linker regions should be removed from EST sequences before submission

EST Files

These are the valid tags and a short description:

TYPE:	Entry type - must be "EST" for EST entries. **Obligatory field**
STATUS:	Status of EST entry - "New" or "Update". **Obligatory field**
CONT_NAME:	Name of contact must be identical string to the contact entry) **Obligatory field**
CITATION:	Journal citation. (Must be identical string to the publication title) Begins on line below tag - use continuation lines if necessary. **Obligatory field**
LIBRARY:	Library name. (Must be identical string to library name entry.) **Obligatory field**
EST#:	EST id assigned by contact lab. For EST updates, this is the string we match on. **Obligatory fiold**
GB#:	GenBank accession number
GB_SEC:	Secondary GenBank accessions
GDB#:	Genome database accession number
GDB_DSEG:	Genome database Dsegment number
CLONE:	Clone id.
SOURCE:	Source providing clone e.g. ATCC
SOURCE_DNA:	Source id number for the clone as pure DNA
SOURCE_INHOST:	Source id number for the clone stored in the host.
OTHER_EST:	Other ESTs on this clone.
DBNAME:	Database name for cross-reference to another database
DBXREF:	Database cross-reference acce-ssion
PCR_F:	Forward PCR primer sequence
PCR_B:	Backward PCR primer sequence
INSERT:	Insert length (in bases)
ERROR:	Estimated error in insert length (bases)
PLATE:	Plate number or code
ROW:	Row number or letter
COLUMN:	Column number or letter
SEQ_PRIMER:	Sequencing primer description or sequence.
P_END:	Which end sequenced e.g. 5'
HIQUAL_START:	Base position of start of highest quality sequence (default=1)
HIQUAL_STOP:	Base position of last base of highest quality sequence.
DNA_TYPE:	cDNA (default), Genomic, Viral, Synthetic, Other
PUBLIC:	Date of public release. Leave blank for immediate re-lease. Format: 9/11/1994 (MM/DD/YYYY) **Obligatory field**
PUT_ID:	Putative identification of se-quence by submitter.
TAG_LIB:	Name of library whose tag is found in this sequence.
TAG_TISSUE:	Tisd, put 'Not found' in this field.
POLYA:	Y or Nsue that was source for the tagged library, if a li-brary tag was found.
TAG_SEQ:	The actual sequence of the library tag found in the EST read. If the tag was searched for and not foun to indicate if a polyA tail was or was not found in the EST sequence.
COMMENT:	Comments about EST. Starts on line below COMMENT: tag.
SEQUENCE:	Sequence string. Starts on line below SEQUENCE: tag. **Obligatory field**

Fig. 2.4: Representation of EST File

- The TYPE field is obligatory at the beginning of each entry, even if there are multiple entries of a given type in a file.
- Valid data values for the EST STATUS field are New (new entry) or Update (change existing EST entry).
- When updating an EST, only the fields with text present in the EST file will be changed. The text in the update file will replace entirely the text for that field in the current version of the file.
- The DNA_TYPE is assumed to be cDNA, so this field may be omitted unless the DNA type differs from this.
- Sequences start on line below the SEQU-ENCE field tag, and should be no more than 60 bases per line with no blank spaces.
- It is very important that the strings in the following fields be completely identical, for purposes of automatic matching:

CONT_NAME field of EST file and NAME field of the Contact file.

LIBRARY field of EST file and NAME field of the Library file.

CITATION field of EST file and TITLE field of the Publication file.

We scan these fields from the EST file and matching them automatically to Library, Contact and Publication records in the other tables, so content, spelling, letter case and spacing must match.

- New fields added April 1999: Four new fields have been added to the EST table: POLYA is used to indicate whether a poly A tail sequence was found in the EST read.

TAG_LIB, TAG_TISSUE, TAG_SEQ: These were added because some researchers, e.g. Bento Soares, are pooling libraries to make subtracted libraries. It is important to many people to know which of the libraries in the pooled library a particular sequence came from, so short sequence tags are being included in the library preps. These tags are specific for each library, so if you find that tag in a cDNA read, you know it came from that library, even though you found it in a library which was a pool of libraries.

TAG_LIB: It contains the name of the library whose tag you found in this sequence, e.g. NCI_CGAP_Pr1. Leave the field blank if you did not find one. If the library name in this field does not match the name of the library the sequence came from, or one of the libraries used to make a subtracted library, this field will alert us that there is a problem.

TAG_TISSUE: It contains the tissue that the tagged library was made from, e.g. prostate in the above example.

TAG_SEQ: It contains the actual tag sequence you found in the EST read. If you did not find one, put "None found" in this field - that will distinguish this case from one where the tag has not been searched for.

KEGG (Kyoto Encyclopedia of Genes and Genomes)

KEGG is a collection of online databases dealing with genomes, enzymatic pathways, and biological chemicals. The PATHWAY database records networks of molecular interactions in the cells, and variants of them specific to particular organisms. The KEGG, the Kyoto Encyclopedia of Genes and Genomes, was initiated by the Japanese human genome programme in 1995 [According to the developers they consider KEGG to be a "computer representation" of the biological system]. The KEGG database can be utilized for modelling and simulation, browsing and retrieval of data. It is a part of the systems biology approach.

KEGG maintains five main databases

1>KEGG 2>Atlas KEGG Pathway 3>KEGG Genes 4>KEGG Ligand 5>KEGG BRITE

Databases

KEGG connects known information on molecular interaction networks, such as pathways and complexes (this is the Pathway Database), information about genes and proteins generated by genome projects (including the gene database) and information about biochemical compounds and reactions (including compound and reaction databases). These databases are different networks, known as the protein network, the gene universe and the chemical universe respectively. There are efforts in progress to add to the knowledge of KEGG, including information regarding ortholog clusters in the KO (KEGG Orthology) database.

KEGG Pathways: *Metabolism *Genetic Information Processing *Environmental Information Processing *Cellular Processes *Human Diseases *Drug development

Ligand Database: *Compound *Drug *Glycan *Reaction *RPAIR (Reactant pair alignments) *Enzyme.

3

Role of Software Engineers in Bioinformatics

The tools of computer science, statistics and mathematics are critical for studying biology as an informational science. Curiously, biology is the only science that at its very heart, employs a digital language. The grand challenge in biology is to determine how the digital language of the chromosomes is converted into 3-D and 4-D (time varying) languages of living organisms.

Need for Software Automation

DNA encodes the information necessary for building and maintaining life. DNA is a non-branching, double-stranded macromolecule in which the nucleotide building blocks (A,C,G,T) are linked. Bases are arranged in A-T and C-G pairs. Small viral genomes of the order of several thousand bases were the first to be sequenced in 1970. Few years later, genomes of the order of 40 kilobase pairs represented the limit of what could reasonably be sequenced. At this stage, the need for automation was recognized and methods were applied to the degree possible. By the year 1997, the yeast genome consisting of 12 Megabase pairs was completed, and in 1998, the conclusion of the 100 Megabase pairs nematode genome project was announced. The 180 Megabase pairs fruit-fly genome was also completed. All of these projects relied on substantially higher levels of software automation. The most ambitious project so far, i.e. sequencing of the 3 Giga base pairs human genome has been completed. For this effort, and those yet to come, software automation lies at the very core of the planning and executing of the project.

The need for automation is driven largely by the trend of handling ever larger sizes of DNA and the corresponding increase in the amount of raw data this entails. Mathematical analysis indicates that the size of a project is roughly proportional to the size of the genome.

This is due to the fact that the amount of information obtained for an individual sequencing experiment is relatively constant and is independent of the genome size. To meet the projected goals, modern large scale sequencing centers have developed throughput capacities of the order of several million experiments per month, with data processing handled on a continuous basis. Managing such large projects without a high degree of automation would clearly be impossible in terms of cost and time requirements.

DNA is the basic genetic material. It transmits hereditary characters from one generation to the next. During synthesis of proteins, mRNA which act as the messengers of information (the exact genetic code) are build from DNA. Proteins are synthesized using mRNA molecules. Protein interactions give rise to information pathways and networks which help in building cells which are identical to their parent cells. Clustering of many cells in a predefined format composes a tissue. An organ is a combination of tissues and an organism is nothing but an organization of organs (Fig. 3.1).

DNA
↓
mRNA
↓
Protein
↓
Protein interactions
↓
Informational pathways
↓
Informational networks
↓
Cells
↓
Tissues
↓
Organs
↓
Individual

Fig. 3.1: Genetic information flow in biological systems

The challenge for computer professionals is to create tools that can capture and integrate these different levels of biological information.

What Challenges Does Biology Offer Computer Scientists?

Computers are powerful devices for understanding any system that can be described in a mathematical way. As our understanding of biological processes has grown and deepened, it is not surprising, then, that the disciplines of computational biology and, more recently, bioinformatics, have evolved from the intersection of classical biology, mathematics, and computer science.

The goal of biology, in the era of the genome projects, is to develop a quantitative understanding of how living things are built from the genome that encodes them. Cracking the genome code is complex. At the very simplest level, we still have difficulty identifying unknown genes by computer analysis of genomic sequence. We still have not managed to predict or model how a chain of amino acids folds into the specific structure of a functional protein. Beyond the single-molecule level, the challenges are immense. The sheer amount of data in GenBank is now growing at an exponential rate, and as datatypes beyond DNA, RNA, and protein sequence begin to undergo the same kind of explosion, simply managing, accessing, and presenting this data to users in an intelligible form is a critical task. Human-computer interaction specialists need to work closely with academic and clinical researchers in the biological sciences to manage such staggering amounts of data. Biological data is very complex and interlinked. A spot on a DNA array, for instance, is connected not only to immediate information about its intensity, but to layers of information about genomic location, DNA sequence, structure, function, and more. Creating information systems that allow biologists to seamlessly follow these links without getting lost in a sea of information is also a huge opportunity for computer scientists.

Finally, each gene in the genome is not an independent entity. Multiple genes interact to form biochemical pathways, which in turn feed into other pathways. Biochemistry is influenced by the external environment, by interaction with pathogens, and by other stimuli. Putting genomic and biochemical data together into quantitative and predictive models of biochemistry and physiology will be the work of a generation of computational biologists. Computer scientists, mathematicians, and statisticians will be a vital part of this effort.

What Skills Should a Bioinformatician Have?

Some of the core requirements for bioinformaticians are:

1. Fairly deep background in some aspects of molecular biology. It can be biochemistry, molecular biology, molecular biophysics, or even molecular modelling.
2. Understanding how and why DNA sequence is transcribed into RNA and translated into protein is vital.
3. Substantial experience with at least one or two major molecular biology software packages, either for sequence analysis or molecular modelling. The experience of learning one of these packages makes it much easier to learn to use other software quickly.
4. One should be comfortable working in a command-line computing environment. Working in Linux or Unix will provide this experience.
5. Experience in programming languages such as C/C++/JAVA, as well as in a scripting language such as Perl or Python.
6. Experience with bioinformatics tools, such as BLAST, BLAT, sequence analysis algorithms and clustering tools.
7. Experience in using bioinformatics resources, such as the UCSC genome browser and Entrez. You will need to be familiar with the National Center for Bioinformatics.
8. Good data visualization skills. You will need to be able to take complex data and interpret it into models and other ways that make it understandable for biologists and other team members.
9. Experience in Database design and implementation, algorithm development.
10. Skill in data mining—being able to extract data from multiple resources is invaluable.

Genetic Algorithms

All that computers can do is implement algorithms. Hence when we talk of using computers for processing of biological information, we have to define precise mathematical algorithms. Following are a few absolutely basic algorithms in bioinformatics.

Database Searching

Database interrogation can take the form of text queries, (e.g. display all the human adrenergic receptors) or sequence similarity searches (e.g. given the sequence of a human adrenergic receptor, display all the similar sequences in the database). Sequence similarity searches are straightforward because the data in the databases is mostly in the form of sequences.

Comparing Two Sequences

Let us take the case of comparing two protein sequences. The alphabet complexity is 20, since a protein is nothing but a sequence of amino acids and there are 20 possible amino acids. The naïve approach is to line up the sequences against each other and insert additional characters to bring the two strings into vertical alignment. More the matches, more is the closeness in the two sequences.

The process of alignment can be measured in terms of the number of gaps introduced and the number of mismatches remaining in the alignment. A metric relating such parameters represents the distance between two sequences.

Multiple Sequence Alignment

In the previous subsection, we saw pairwise sequence alignment, which is fundamental to sequence analysis. However, analysis of groups of sequences that form gene families requires the ability to make connections between more than two members of the group, in order to reveal subtle conserved family characteristics. The goal of multiple sequence alignment is to generate a concise information-rich summary of sequence data in order to inform decision-making on the relatedness of sequences to a gene family.

Multiple sequence alignment is a 2D table, in which the rows represent individual sequences and the columns the residue positions. The sequences are laid on to this grid in such a manner that: (a) the relative positioning of residues within any one sequence is preserved, and (b) similar residues in all the sequences are brought into vertical register.

Bioinformatics and Computational Biology

Bioinformatics support is essential to the implementation of genome projects and for public access to their output. Bioinformatics needs for the genome project fall into two broad areas: (i) databases and (ii) development of analytical tools. Collection, analysis, annotation, and storage of the ever increasing amounts of mapping, sequencing, and expression data in publicly accessible, user-friendly databases is critical to the project's success. In addition, the community needs computational methods that will allow scientists to extract, view, annotate, and analyze genomic information efficiently. Thus, the genome project must continue to invest substantially in these areas. Conservation of resources through development of portable software should be encouraged.

a. Improve content and utility of databases. Databases are the ultimate repository of genome project's data. As new kinds of data are generated and new biological relationships discovered, databases must provide for continuous and rapid expansion and adaptation to the evolving needs of the scientific community. To encourage broad use, databases should be responsive to a diverse range of users with respect to data display, data deposition, data access, and data analysis. Databases should be structured to allow the queries of greatest interest to the community to be answered in a seamless way. Communication among databases must be improved. Achieving this will require standardization of nomenclature. A database of human genomic information, analogous to the model organism databases and including links to many types of phenotypic information, is needed.
b. Develop better tools for data generation, capture, and annotation. Large-scale, high-throughput genomic centers need readily available, transportable informatics tools for commonly performed tasks such as sample tracking, process management, map generation, sequence finishing, and primary annotation of data. Smaller users urgently need reliable tools to meet their sequencing and sequence analysis needs. Readily accessible information about the availability and utility of various tools should be provided, as well as training in the use of tools.
c. Develop and improve tools and databases for comprehensive functional studies. Massive amounts of data on gene expression and function will be generated in the near future. Databases that can organize and display this data in useful ways need to be developed. New statistical and mathematical methods are needed for analysis and comparison of expression and function data, in a variety of cells and tissues, at various times and under different conditions. Also needed are tools for modelling complex networks and interactions.

d. Develop and improve tools for representing and analyzing sequence similarity and variation. The study of sequence similarity and variation within and among species will become an increasingly important approach to biological problems. There will be many forms of sequence variation, of which SNPs will be only one type. Tools need to be created for capturing, displaying, and analyzing information about sequence variation.
e. Create mechanisms to support effective approaches for producing robust, exportable software that can be widely shared. Many useful software products are being developed in both academia and industry that could be of great benefit to the community. However, these tools generally are not robust enough to make them easily exportable to another laboratory. Mechanisms are needed for supporting the validation and development of such tools into products that can be readily shared and for providing training in the use of these products. Participation by the private sector is strongly encouraged.

Application of Perl in Biology

Perl is a popular programming language that is extensively used in areas such as bioinformatics and web programming. Perl has become popular with biologists because it is so well-suited to several bioinformatics tasks.

Perl is a programming language that was created in 1986 as the result of one man's frustration with a task involved generating reports from a lot of text files with cross-references. Being tired of writing specific utilities to manage similar tasks dealing with extracting information from text and generating reports, Larry Wall invented a new language and wrote an interpreter in C for it that later became known as Perl (stands for practical extraction report language). The initial emphasis of Perl was on system management and text handling. More features were added during later revisions, including regular expressions, signals, and network sockets.

Reasons of Popularity of Perl in Biology

The Perl programming language is probably the most commonly used language in bioinformatics, partly for the following reasons. First, Perl is less structured than traditional programming languages, and it is easy to learn for biologists. Perl takes care of many of the low-level tasks in traditional programming, such as memory allocation, and developers can concentrate on solving the problem at hand. Perl is interpreted, which enables the quick and dirty creation of portable (i.e. platform-independent) analysis programs. Perl is also free software and available for a variety of operating systems, including Windows and Mac, as well as Linux and Unix platforms. It is often the case that a problem can be solved using Perl programming with a few lines of code while requiring several times more lines of code in C or Java.

Application of Java in Biology

Another popular language used in biology is Java, which was developed by Sun Microsystems. Java is an object-oriented language similar to C++, but simplified to eliminate language features that cause common program errors. Similar to Perl, Java is an interpreted language. Java source code files (files with a .java extension) are first compiled into a format called bytecode (files with a class extension), which can then be executed by a Java interpreter. Compiled Java code can run on most computers because Java interpreters and runtime environments, known as Java Virtual Machines (JVMs), exist for most operating systems, including Windows, Linux/Unix, and Mac. Bytecode can also be converted directly into executable applications using a just-in-time compiler (JIT).

Reasons of Popularity of Java in Biology

Traditionally, the language of choice for bioinformaticians has been Perl. Perl scripts usually require prerequisite dependency installations, and they lack the dynamic graphical user interface (GUI) interactions inherent in Java. Additionally, the value of stand-alone bioinformatics applications created with Perl is limited in scope and in its contribution back to research biologists. For this reason, bioinformaticians have been using Java to deliver applications to researchers at all levels of computational ability and provide dynamic GUI interactions.

Another feature making Java popular is its object-oriented property and a rich set of existing Java application programming interface (API) packages for bioinformatics. Object-oriented programming defines not only the data type of a data structure, but also the types of operations that can be applied to the data structure. In this way, the data structure becomes an object that includes both data and functions. In addition, programmers can create relationships between one object and another. For example, objects can inherit characteristics from other objects. One of the principal advantages of object-oriented pr ogramming techniques is that they enable programmers to create modules that do not need to be changed when a new type of object is added. A programmer can simply create a new object that inherits many of its features from existing objects. This makes object-oriented programs easier to modify.

4

Sequence Alignment

In bioinformatics, a sequence alignment is a way of arranging the sequences of DNA, RNA, or protein to identify regions of similarity that may be a consequence of functional, structural, or evolutionary relationships between the sequences. Aligned sequences of nucleotide or amino acid residues are typically represented as rows within a matrix. Gaps are inserted between the residues so that identical or similar characters are aligned in successive columns.

Sequence Alignment and Similarity

The idea of aligning two sequences (of possibly different sizes) is to write one on top of the other, and break them into smaller pieces by inserting spaces in one or the other so that identical subsequences are eventually aligned in a one-to-one correspondence—naturally, spaces are not inserted in both sequences at the same position. In the end, the sequences end up with the same size.

The following example illustrates an alignment between the sequences A= .ACAAGACAGCGT. and B=.AGAACAAGGCGT.

```
A = A C A A G A C A G - C G T
    |   | |   |   | |   | | |
B = A G A A C A - A G G C G T
```

Fig. 4.1: Alignment of two sequences

The objective is to match identical subsequences as far as possible. In the example, nine matches are highlighted with vertical bars. However, if the sequences are not identical, mismatches are likely to occur as different letters are aligned together. Two mismatches can be identified in the example: a "C" of A aligned with a "G" of B, and a "G" of A aligned with a "C" of B.

The insertion of spaces produced gaps in the sequences. They were important to allow a good alignment between the last three characters of both sequences.

An alignment can be seen as a way of transforming one sequence into the other. From this point of view, a mismatch is regarded as a substitution of characters. A gap in the first sequence is considered an insertion of a character from the second sequence into the first one, whereas a gap in the second sequence is considered a deletion of a character of the first sequence. In the previous example, A can be converted into B in four steps: (1) substitute the first "C" for a "G"; (2) substitute the first "G"for a "C"; (3) delete the second "C"; and (4) insert a "G" before the last three characters.

Once the alignment is produced, a score can be assigned to each pair of aligned letters, called aligned pair, according to a chosen scoring scheme. We usually reward matches and penalize mismatches and gaps. The overall score of the alignment can then be computed by adding up the score of each pair of letters. For instance, using a scoring scheme that gives a +1 value to matches and 1 to mismatches and gaps, the alignment of Figure 1 scores 9 • (1) + 2 • (1) + 2 • (1) = 5.

The similarity of two sequences can be defined as the best score among all possible alignments between them. Note that it depends on the choice of scoring scheme. In the next sections, the problem of finding the best alignment of two sequences (an alignment that gives the highest score) will be addressed. A related notion is that of distance. However, this work will focus on similarity, as it is the preferred choice for biological applications.

Thus far, this section has described a type of alignment known as global alignment since we are interested in the best match covering the two sequences in their entirety. Frequently, though, biologists are interested in short regions of local similarity. A local alignment is one that looks for best alignments between "pieces", or more precisely, substrings of both sequences.

Substitution Matrices

However, biologists frequently use scoring schemes that take into account physicochemical properties or evolutionary knowledge of the sequences being aligned. This is common when protein sequences are compared. For instance, for some reason one might want to penalize the mismatch of an aspartic acid (D) with leucine (L) more heavily than a mismatch between the same aspartic acid with, say, histidine (H). Similarly, one may want to reward a match of two cysteines (C) better than two alanines (A).

This type of scoring schemes is called alphabet-weight scoring schemes, and is usually implemented by a substitution matrix. Currently, two types of amino acid substitution matrices are being largely used by biologists for practical protein sequence alignment: PAM and BLOSUM. They were developed from different concepts but have the same structure. In fact, they are a series of matrices with varied degrees of sensibility. Substitution matrices allow for the possibility of giving a positive score for a mismatch, what is sometimes called an approximate or partial match.

The PAM Matrices (Acronym for Point Accepted Mutations)

PAM matrices were proposed by Dayhoff et al. in 1978 based on observations of hundreds of alignments of closely related proteins. The frequencies of substitution of each pair of amino acids were extracted from alignments of proteins of small evolutionary distance, below 1% divergence, i.e. at most one mutation per 100 amino acids, on average. These frequencies, normalized to account for the frequencies of random occurrences of single amino acids, resulted in the PAM-1 probability transition matrix. The PAM-1 matrix reflects an amount of evolutionary change that yields on average one mutation per 100 amino acids. Accordingly, it is suitable for comparison of proteins which have diverged by 1% or less. The acronym PAM stands for percent of accepted mutations (and hence the distance is in percentage) or for point accepted mutations (and hence the distance in number of mutations per 100 amino acids).

The PAM-1 matrix is then extrapolated to yield the family of PAM-k matrices. Each PAM-k matrix is obtained from PAM-1 by k consecutive multiplication, and is suitable for comparison of sequences which have diverged k%, or are k evolutionary units apart. For example, PAM-250 = (PAM-1)250 reflects the frequencies of mutations for proteins which have diverged 250% (250 mutations per 100 amino acids). The actual scoring matrices that are used by search programs are derived from the transition probability matrices and the background probabilities. The score of each pair s(a,b) is defined as the logarithm of the likelihood ratio of the transition probability Mab (mutation) versus the probability of a random occurrence of the amino acid b in the second sequence, i.e. s(a,b) = logMab/Pb.

The BLOSUM family of scoring matrices

Unlike PAM matrices, which are extrapolated from a single matrix PAM-1, the BLOSUM series of matrices was constructed by direct observation of sequence alignments of related proteins, at different levels of sequence divergence. The matrices are based on 'blocks'—a collection of multiple alignments of similar segments without gaps, each block representing a conserved region of a protein family. These blocks provide a list of (accepted) substitutions, and a log odds scoring matrix can be defined based on the observed relative frequency of aligned pairs of amino acids q_{ab}, and the expected probability of pairs e_{ab} estimated from the population of all observed pairs

$$s_{ab} = \frac{\log q_{ab}}{e_{ab}}$$

To reduce the bias in the amino acid pair frequencies caused by multiple counts from closely related sequences, segments in a block with at least x% identity are clustered and pairs are counted between clusters, i.e. pairs are counted only between segments less than x% identical. When counting pairs frequencies between clusters, the contributions of all segments within a cluster are averaged, so that each cluster is weighted as a single sequence. Varying the percentage of identity within clusters results in a family of matrices BLOSUM- , where ranges from 30 to 100.

For example, BLOSUM-62 is based on pairs that were counted only between segments less than 62% identical.

Choosing the scoring matrix

When comparing two sequences, the most effective matrix to use is the one which corresponds to the evolutionary distance between them. However, we usually do not know this distance. Therefore, it is recommended to use several scoring matrices which cover a range of evolutionary distances, e.g. PAM-40, PAM-120, and PAM-250. In general, low PAM matrices are well suited to finding short but strong similarities, while high PAM matrices are best for finding long regions of weak similarity.

Standard Algorithms

In the previous section, the similarity of two sequences was defined as the best score among all possible alignments between them. Examining every alternative can be virtually impossible unless the sequences are relatively short, because the number of potential alignments is exponential. Fortunately, there are simpler solutions that yield much more efficient algorithms.

Needleman-Wunsch

The standard global alignment algorithm, referred to as Needleman-Wunsch after its original authors, computes the similarity between two sequences A and B of lengths m and n, respectively, using a dynamic programming approach. Dynamic programming is a strategy of building a solution gradually using simple recurrences. The key observation for the alignment problem is that the similarity between sequences A[1..n] and B[1..m] can be computed by taking the maximum of the three following values:

- The similarity of A [1..n 1] and B [1..m 1] plus the score of substituting A [n] for B [m];

- The similarity of A[1..n 1] and B[1..m] plus the score of deleting aligning A[n];
- The similarity of A[1..n] and B[1..m 1] plus the score of inserting B[m].

From this observation, the following recurrence can be derived:

sim (A[1..i], B[1..j]) = max {
  sim (A[1..i 1], B[1..j 1]) + sub (A[i], B[j]);
  sim (A[1..i 1], B[1..j]) + del (A[i]);
  sim (A[1..i], B[1..j 1]) + ins (B[j]) }

where sim (A, B) is a function that gives the similarity of two sequences A and B, and sub (a, b), del (c) and ins (c) are scoring functions that give the score of a substitution of character a for character b, a deletion of character c, and an insertion of character c, respectively. This recurrence is complete with the following base case:

sim (A[0], B[0]) = 0;

where, A[0] and B[0] are defined as empty strings.

To solve the problem with this recurrence, the algorithm builds an (n +1) (m +1) matrix M where each M[i, j] represents the similarity between sequences A[1..i] and B[1..j] (Fig. 4.2).

The first row and the first column represent alignments of one sequence with spaces. M[0, 0] represents the alignment of two empty strings, and is set to zero. All other entries are computed with the following formula:

M[i, j] = max { M[i 1, j 1] + sub (A[i], B[j]);
  M[i 1, j] + del (A[i]);
  M [i, j 1] + ins (B[j]) }

The matrix can be computed either row by row (left to right) or column by column (top to bottom). In the end, M[n, m] will contain the similarity score of the two sequences. Since there are (m+1) • (n+1) positions to compute and each take a constant amount of work, this algorithm has time complexity of O(n2). Clearly, it has also quadratic space complexity since it needs to keep the entire matrix in memory.

Once the matrix has been computed, the actual alignment can be retrieved by tracing a path in the matrix from the last position to the first (Fig. 4.2). The trace is a simple procedure that compares the value at each M[i, j] to the values of its left, top and diagonal entries according to the formula given above. For instance, if M[i, j] = M [i, j 1] + ins (B[j]), the trace reports an insertion of character B[j] and proceeds to entry M[i, j 1]. Alternatively, pointers can be saved on each entry during the computation of the matrix so that this evaluation step can be avoided at the cost of more memory usage. Since the path can be as long as O(m+n), this procedure has linear time complexity. Note that sometimes more than one path can be traversed and, as a result, multiple high-scoring alignments can be produced. In the matrix of Fig. 4.2, two optimal alignments can be retrieved (Figs 4.1 and 4.3).

		0	1	2	3	4	5	6	7	8	9	10	11	12
		-	A	G	A	A	C	A	A	G	G	C	G	T
0	-	0	-1	-2	-3	-4	-5	-6	-7	-8	-9	-10	-11	-12
1	A	-1	1	0	-1	-2	-3	-4	-5	-6	-7	-8	-9	-10
2	C	-2	0	0	-1	-2	-1	-2	-3	-4	-5	-6	-7	-8
3	A	-3	-1	-1	1	0	-1	0	-1	-2	-3	-4	-5	-6
4	A	-4	-2	-2	0	2	1	0	1	0	-1	-2	-3	-4
5	G	-5	-3	-1	-1	1	1	0	0	2	1	0	-1	-2
6	A	-6	-4	-2	0	0	0	2	1	1	1	0	-1	0
7	C	-7	-5	-3	-1	-1	1	1	1	0	0	2	1	0
8	A	-8	-6	-4	-2	0	0	2	2	1	0	1	1	2
9	G	-9	-7	-5	-3	-1	-1	1	1	3	2	1	2	1
10	C	-10	-8	-6	-4	-2	0	0	0	2	2	3	2	1
11	G	-11	-9	-7	-5	-3	-1	-1	-1	1	3	2	4	3
12	T	-12	-10	-8	-6	-4	-2	0	0	0	2	2	3	5

Fig. 4.2: Standard dynamic programming matrix for the global alignment of sequences A=.ACAAGACAGCGT. and B=.AGAACAAGGCGT. with paths to retrieve optimal alignments indicated with arrows

A = A C A A G A C A – G C G T
| | | | | | | | |
B = A G A A C A – A G G C G T

Fig. 4.3: Another optimal alignment retrievable from the matrix of Fig. 4.2.

It is often useful to see the dynamic programming solution for the sequence alignment problem as a directed weighted graph with (n +1) (m +1) nodes representing each entry (i, j) of the matrix, and having the following edges:

- ((i 1, j 1), (i, j)) with weight equals to sub (A[i], B[j]);
- ((i 1, j), (i, j)) with weight equals to del (A[i]);
- ((i, j 1), (i, j)) with weight equals to ins (B[j]);

A path from node (0, 0) to (n, m) in the alignment graph corresponds to an alignment between the two sequences, and the problem of retrieving an optimal alignment is converted to the problem of finding a path in the graph with highest weight.

Smith-Waterman

The Smith-Waterman algorithm is a well-known algorithm for performing local sequence alignment; that is, for determining similar regions between two nucleotide or protein sequences. Instead of looking at the total sequence, the Smith-Waterman algorithm compares segments of all possible lengths and optimizes the similarity measure.

The algorithm was first proposed by Temple Smith and Michael Waterman in 1981. Like the Needleman-Wunsch algorithm, of which it is a variation, Smith-Waterman is a dynamic programming algorithm. As such, it has the desirable property that it is guaranteed to find the optimal local alignment with respect to the scoring system being used (which includes the substitution matrix and the gap-scoring scheme). The main difference to the Needleman-Wunsch algorithm is that M[i, j] contains the similarity between suffixes of A[1..i] and B[1..j]. As a result, the recurrence relation is slightly altered because an empty string is a suffix of any sequence and, therefore, a score of zero is always possible. The formula for computing M[i, j] becomes:

M[i, j] = max { 0;
M[i 1, j 1] + sub (A[i], B[j]);
M[i 1, j] + del (A[i]);
M[i, j 1] + ins (B[j]) }

Another important distinction is that the score of the best local alignment is the highest value found anywhere in the matrix. This position is the starting point for retrieving an optimal alignment using the same procedure described for the global alignment case. The path ends, however, as soon an entry with score zero is reached. It is trivial to see that the Smith-Waterman algorithm has the same time and space complexity as the Needleman-Wunsch 4

A Note on Gap Penalty Functions

The alignments algorithms described in previous sections assume that gap costs are given by a constant gap penalty function. This means that a gap is given the same score no matter how long it is. However, biologists have long recognised that insertions and deletions generally do not occur a single base at a time. Therefore, when biomolecular sequences are compared, it is commonly accepted that a gap of k spaces is more likely to appear than k incidences of a single gap spread across the sequences. In order to make this distinction, a general gap penalty function is needed so that the cost of a gap is a function of its length. Unfortunately, if a general gap penalty function is required, it is not possible to compute an alignment with the algorithms described earlier because the scoring scheme is no longer additive. Moreover, the solution to such a problem has complexity $O(n^3)$.

A more restrictive function, called affine gap penalty function, allows an $O(n^2)$ solution at the same time that it is still able to distinguish between a gap of k spaces and k isolated gaps by penalising two events independently, the opening and the extension of a gap. It is important to observe that a constant gap penalty function is as a special case of an affine function in which the cost of opening a gap is zero. Although the affine model is probably the most common function used in practice, this work is focused on algorithms with constant gap penalty functions.

Tools used for Alignment of Sequences

BLAST

Basic local alignment search tool, or BLAST, is an algorithm for comparing primary biological sequence information, such as the amino acid sequences of different proteins or the nucleotides of DNA sequences. A BLAST search enables a researcher to compare a query sequence with a library or database of sequences, and identify library sequences that resemble the query sequence above a certain threshold. For example, following the discovery of a previously unknown gene in the mouse, a scientist will typically perform a BLAST search of the human genome to see if humans carry a similar gene; BLAST will identify sequences in the human genome that resemble the mouse gene based on similarity of sequence. The BLAST program was designed by Eugene Myers, Stephen Altschul, Warren Gish, David J. Lipman and Webb Miller at the NIH and was published in *J. Mol. Biol.* in 1990.

To run, BLAST requires a query sequence to search for, and a sequence to search against (also called the target sequence) or a sequence database containing multiple such sequences. BLAST will find subsequences in the database which are similar to subsequences in the query. In typical usage, the query sequence is much smaller than the database, e.g. the query may be one thousand nucleotides while the database is several billion nucleotides.

The main idea of BLAST is that there are often high-scoring segment pairs (HSP) contained in a statistically significant alignment. BLAST searches for high scoring sequence alignments between the query sequence and sequences in the database using a heuristic approach that approximates the Smith-Waterman algorithm. The exhaustive Smith-Waterman approach is too slow for searching large genomic databases such as GenBank. Therefore, the BLAST algorithm uses a heuristic approach that is less accurate than the Smith-Waterman but over 50 times faster. The speed and relatively good accuracy of BLAST are among the key technical innovation of the BLAST programs.

Variants

BLAST is a family of programs that includes BLASTN, BLASTP, BLASTX TBLASTN,and TBLASTX. BLASTN queries nucleotide sequences with a nucleotide sequence database. BLASTP uses protein sequences as queries to search against a protein sequence database. BLASTX uses nucleotide sequences as queries and translates them in all six reading frames to produce translated protein sequences, which are used to query a protein sequence database. TBLASTN queries protein sequences to a nucleotide sequence database with the sequences translated in all six reading frames. TBLASTX uses nucleotide sequences, which are translated in all six frames, to search against a nucleotide sequence database that has all the sequences translated in six frames. In addition, there is also a bl2seq program that performs local alignment of two user-provided input sequences.

The graphical output includes horizontal bars and a diagonal in a two-dimensional diagram showing the overall extent of matching between the two sequences.

The BLASTweb server(www.ncbi.nlm.nih.gov/BLAST/) has been designed in such a way as to simplify the task of program selection. The programs are organized based on the type of query sequences, protein sequences, nucleotide sequences, or nucleotide sequence to be translated.

The choice of the type of sequences also influences the sensitivity of the search. Generally speaking, there is a clear advantage of using protein sequences in detecting homologs. This is because DNA sequences only comprise four nucleotides, whereas protein sequences contain 20 amino acids. This means that there is at least a fivefold increase in statistical complexity for protein sequences. More importantly, amino acid substitution matrices incorporate subtle differences in physicochemical properties between amino acids, meaning that protein sequences are far more informative and sensitive in detection of homologs. This is why searches using protein sequences can yield more significant matches than using DNA sequences. For that reason, if the input sequence is a protein-encoding DNA sequence, it is preferable to use BLASTX, which translates it in six open reading frames before sequence comparisons are carried out.

If one is looking for protein homologs encoded in newly sequenced genomes, one may use TBLASTN, which translates nucleotide database sequences in all six open reading frames. This may help to identify protein coding genes that have not yet been annotated.If a DNA sequence is to be used as the query, a protein-level comparison can be done with TBLASTX. However, both programs are very computationally intensive and the search process can be very slow.

Statistical Significance

The BLAST output provides a list of pairwise sequence matches ranked by statistical significance. The significance scores help to distinguish evolutionarily related sequences from unrelated ones. Generally, only hits above a certain threshold are displayed. Deriving the statistical measure is slightly different from that for single pairwise sequence alignment; the larger the database, the more unrelated sequence alignments there are. This necessitates a new parameter that takes into account the total number of sequence alignments conducted, which is proportional to the size of the database. In BLAST searches, this statistical indicator is known as the E-value (expectation value), and it indicates the probability that the resulting alignments from a database search are caused by random chance. The E-value is related to the P-value used to assess significance of single pairwise alignment. BLAST compares a query sequence against all database sequences, and so the E-value is determined by the following formula:

$$E = m \quad n \quad P$$

where m is the total number of residues in a database, n is the number of residues in the query sequence, and P is the probability that an HSP alignment is a result of random chance. For example, aligning a query sequence of 100 residues to a database containing a total of 10^{12} residues results in a P-value for the ungapped HSP region in one of the database matches of $1 \quad 1^{20}$. The E-value, which is the product of the three values, is $100 \quad 10^{12} \quad 10^{20}$, which equals 10^{6}. It is expressed as 1e ^{6} in BLAST output. This indicates that the probability of this database sequence match occurring due to random chance is 10^{6}.

The E-value provides information about the likelihood that a given sequence match is purely by chance. The lower the E-value, the less likely the database match is a result of random chance and therefore the more significant the match is. Empirical interpretation of the E-value is as follows. If E < 1e ^{50} (or $1 \quad 10^{50}$), there should be an extremely high confidence that the database match is a result of homologous relationships. If E is between 0.01 and 1e ^{50}, the match can be considered a result of homology. If E is between 0.01 and 10, the match is considered not significant, but may hint at a tentative remote homology relationship. Additional evidence is needed to confirm the tentative relationship. If $E > 10$, the sequences under consideration are either unrelated or related by extremely distant relationships that fall below the limit of detection with the current method.

Because the E-value is proportionally affected by the database size, an obvious problem is that as the database grows, the E-value for a given sequence match also increases. Because the genuine evolutionary relationship between the two sequences remains constant, the decrease in credibility of the sequence match as the database grows means that one may "lose" previously detected homologs as the database enlarges. Thus, an alternative to E-value calculations is needed.

A bit score is another prominent statistical indicator used in addition to the E value in a BLAST output. The bit score measures sequence similarity independent of query sequence length and database size and is normalized based on the raw pairwise alignment score. The bit score (S′) is determined by the following formula:

$$S' = (\lambda \quad S \quad \ln K) / \ln 2$$

where λ is the Gumble distribution constant, S is the raw alignment score, and K is a constant associated with the scoring matrix used. Clearly, the bit score (S) is linearly related to the raw alignment score (S). Thus, the higher the bit score, the more highly significant the match is. The bit score provides a constant statistical indicator for searching different databases of different sizes or for searching the same database at different times as the database enlarges.

FASTA

FASTA (FAST ALL, www.ebi.ac.uk/fasta33/) was in fact the first database similarity search tool developed, preceding the development of BLAST. FASTA uses a "hashing" strategy to find matches for a short stretch of identical residues with a length of k. The string of residues is known as ktuples or ktups, which are equivalent to words in BLAST, but are normally shorter than the words. Typically, a ktup is composed of two residues for protein sequences and six residues for DNA sequences.

The first step in FASTA alignment is to identify ktups between two sequences by using the hashing strategy. This strategy works by constructing a lookup table that shows the position of each ktup for the two sequences under consideration. The positional difference for each word between the two sequences is obtained by subtracting the position of the first sequence from that of the second sequence and is expressed as the offset. The ktups that have the same offset values are then linked to reveal a contiguous identical sequence region that corresponds to a stretch of diagonal in a two-dimensional matrix.

The second step is to narrow down the high similarity regions between the two sequences. Normally, many diagonals between the two sequences can be identified in the hashing step. The top ten regions with the highest density of diagonals are identified as high similarity regions. The diagonals in these regions are scored using a substitution matrix. Neighboring high-scoring segments along the same diagonal are selected and joined to form a single alignment. This step allows introducing gaps between the diagonals while applying gap penalties. The score of the gapped alignment is calculated again. In step 3, the gapped alignment is refined further using the Smith–Waterman algorithm to produce a final alignment. The last step is to perform a statistical evaluation of the final alignment as in BLAST, which produces the E-value (Fig. 4.4).

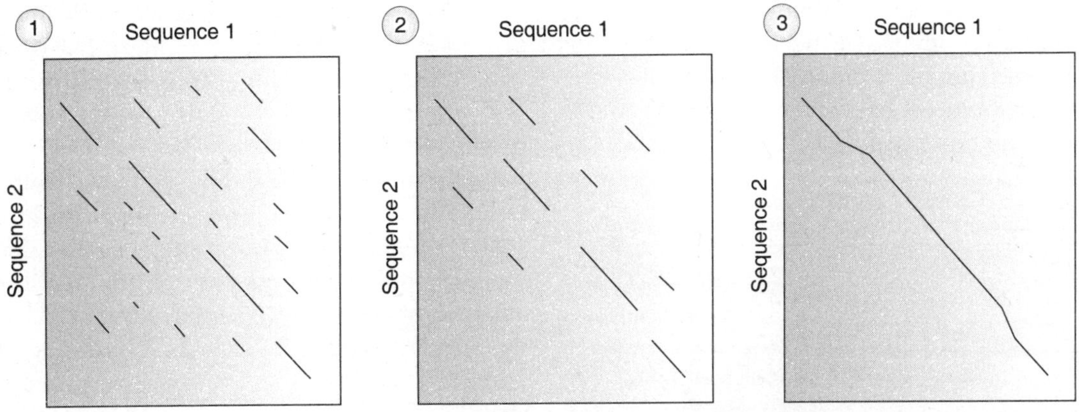

Fig. 4.4: Steps of the FASTA alignment procedure. In step 1 (left), all possible ungapped alignments are found between two sequences with the hashing method. In step 2 (middle), the alignments are scored according to a particular scoring matrix. Only the ten best alignments are selected. In step 3 (right), the alignments in the same diagonal are selected and joined to form a single gapped alignment,which is optimized using the dynamic programming approach

Statistical Significance

FASTA also uses E-values and bit scores. Estimation of the two parameters in FASTA is essentially the same as in BLAST. However, the FASTA output provides one more statistical parameter, the Z-score. This describes the number of standard deviations from the mean score for the database search. Because most of the alignments with the query sequence are with unrelated sequences, the higher the Z-score for a reported match, the further away from the mean of the score distribution, hence, the more significant the match. For a Z-score > 15, the match can be considered extremely significant, with certainty of a homologous relationship. If Z is in the range of 5 to 15, the sequence pair can be described as highly probable homologs. If Z < 5, their relationships is described as less certain.

Table 4.1: Free pairwise sequence comparison programs over the internet

Name	*Address*	*Alignment type*
Lalign	www.ch.embnet.org/software/LALIGN_form.html	Global/Local
USC	www-hto.usc.edu/software/seqaln/seqaln-query.html	Global/Local
Alion	Fold.stanford.edu/alion/	Global/Local
Align	Genome.cs.mtu.edu/align.html	Global/Local
Align	www.ebi.ac.uk/emboss/align	Global/Local
xenAliTwo	www.soe.ucsc.edu/~kent/xenoAli/xenaliTwo.html	Local for DNA
Blast2seqs	www.ncbi.nlm.nih.gov/blast/bl2seq/wblast2.cgi	Local BLAST
Protal2dna	Bioweb.pasteur.fr/seqanal/interfaces/protal2dna.html	Protein against DNA
Pal2nal	Coot.embl.de/pal2nal	Protein against DNA

Generating an alignment is not always enough. You may also need to visualize this alignment and to evaluate its statistical significance. Table 4.2 lists a few sites that enable you to do this online.

Table 4.2: Tools used to visualize alignments

Name	*Address*	*Function*
Lalnview	www.expasy.ch/tools/sim-prot.html	Visualization
Prss	www.ch.embnet.org/software/PRSS_form.html	Evaluation
Graph-align	www.darwin.nmsu.edu/cgi-bin/graph_align.cgi	Evaluation

Multiple Sequence Alignment

A multiple sequence alignment (MSA) is a sequence alignment of three or more biological sequences, generally protein, DNA, or RNA. In many cases, the input set of query sequences are assumed to have an evolutionary relationship by which they share a lineage and are descended from a common ancestor. From the resulting MSA, sequence homology can be inferred and phylogenetic analysis can be conducted to assess the sequences' shared evolutionary origins. Multiple sequence alignment is often used to assess sequence conservation of protein domains, tertiary and secondary structures, and even individual amino acids or nucleotides.

Note, however, that sometimes homology is not assumed between aligned sequences. In such cases, residues aligned in the same column are typically assumed to be structurally equivalent.

Multiple sequence alignment also refers to the process of aligning such a sequence set. Because three or more sequences of biologically relevant length can be difficult and are almost always time-consuming to align by hand, computational algorithms are used to produce and analyze the alignments. MSAs require more sophisticated methodologies than pairwise alignment because they are more computationally complex to produce. Most multiple sequence alignment programs use heuristic methods rather than global optimization because identifying the optimal alignment between more than a few sequences of moderate length is prohibitively computationally expensive.

Dynamic programming and computational complexity

The most direct method for producing an MSA uses the dynamic programming technique to identify the globally optimal alignment solution. For proteins, this method usually involves two sets of parameters: a gap penalty and a substitution matrix assigning scores or probabilities to the alignment of each possible pair of amino acids based on the similarity of the amino acids' chemical properties and the evolutionary probability of the mutation. For nucleotide sequences, a similar gap penalty is used, but a much simpler substitution matrix, wherein only identical matches and mismatches are considered, is typical. The scores in the substitution matrix may be either all positive or a mix of positive and negative in the case of a global alignment, but must be both positive and negative, in the case of a local alignment. In the latter case, it is essential that the average score be less than 0.

For n individual sequences, the naive method requires constructing the n-dimensional equivalent of the matrix formed in standard pairwise sequence alignment. The search space thus increases exponentially with increasing n and is also strongly dependent on sequence length.

To find the global optimum for n sequences, this way has been shown to be an NP-complete problem in 1989, Altschul introduced a practical method that uses pairwise alignments to constrain the n-dimensional search space. In this approach, pairwise dynamic programming alignments are performed on each pair of sequences in the query set, and only the space near the n-dimensional intersection of these alignments is searched for the n-way alignment. The MSA program optimizes the sum of all of the pairs of characters at each position in the alignment (the so-called sum of pair score) and has been implemented in the MSA program. But MSA is still impractical for many multiple sequence alignment applications that require the simultaneous alignment of more than about 20 sequences.

Progressive Alignment Construction

The most widely used approach to multiple sequence alignments uses a heuristic search known as progressive technique (also known as the hierarchical or tree method), that builds up a final MSA by combining pairwise alignments beginning with the most similar pair and progressing to the most distantly related. All progressive alignment methods require two stages: a first stage in which the relationships between the sequences are represented as a tree, called a guide tree, and a second step in which the MSA is built by adding the sequences sequentially to the growing MSA according to the guide tree. The initial guide tree is determined by an efficient clustering method such as neighbor-joining or UPGMA, and may use distances based on the number of identical two letter sub-sequences (as in FASTA rather than a dynamic programming alignment).

Progressive alignment methods are efficient enough to implement on a large scale for many (100s to 1000s) sequences. Progressive alignment services are commonly available on publicly accessible web servers so users need not locally install the applications of interest.

The most popular progressive alignment method has been the Clustal family, especially the weighted variant Clustal W to which access is provided by a large number of web portals including GenomeNet, EBI, and EMBNet.

Another common progressive alignment method called T-Coffee is slower than Clustal and its derivatives but generally produces more accurate alignments for distantly related sequence sets. T-Coffee calculates pairwise alignments by combining the direct alignment of the pair with indirect alignments that aligns each sequence of the pair to a third sequence. It uses the output from Clustal as well as another local alignment program LALIGN, which finds multiple regions of local alignment between two sequences. The resulting alignment and phylogenetic tree are used as a guide to produce new and more accurate weighting factors.

Because progressive methods are heuristics that are not guaranteed to converge to a global optimum, alignment quality can be difficult to evaluate and their true biological significance can be obscure. A 2006, semi-progressive method that improves alignment quality and does not use a lossy heuristic while still running in polynomial time has been implemented in the program PSAlign.

Clustal

ClustalW is the latest of the Clustal software series. Clustal was the first multiple sequence alignment program that could run on almost any platform. Legend has it that Des Higgins made the original design for Clustal on the back of an envelope, in a smoky Dublin pub, in the early 1990s. A few years later, when Des moved to the European Molecular Biology Laboratory in Heidelberg, he took Clustal along — and kept it alive, turning it into one of the most successful series of bioinformatics programs: ClustalV, ClustalW, and ClustalX, an X-window implementation.

ClustalW uses a progressive algorithm. This means that it builds the alignment progressively, adding sequences little by little until the complete multiple sequence alignment is finished. The reason for doing this is that state-of-the-art sequence-alignment programs find it hard to align more than two sequences at a time. Since these pairwise-alignment programs are the only tools we have got on hand to produce a multiple alignment, the only solution is to cheat a little. This is what the progressive algorithm does.

Protocol for using Clustal W

1. Point your browser to the EBI ClustalW server page at www.ebi.ac.uk/clustalw.
2. Paste the sequences you collected in the sequence window.
3. Choose fasta from the alignment pull-down menu.
4. Use the output format pull-down menu to set the selection of your choice.
   It is safe to use Aln without numbers, the default ClustalW format.
   It is never too late to change a format.
5. Choose input from the output order pull-down menu.

Here are a few things to remember about setting output:

- When you set the output order to input, ClustalW outputs your sequences in their original order.
- If you set the output order to aligned, the sequences appear in the order they were aligned. This order depends on the guide tree, and it puts closely related sequences next to each other.
- Although the Aligned output is more informative than input, aligned output makes it difficult to compare alignments generated with different methods or different parameters. Usually it is better to prearrange your sequences in the most informative way and then select the input option.

6. Do not select a tree type: No need to select a tree type here. Any tree computed would be estimated from the unaligned sequences. If you want to compute a tree, compute the alignment first and then turn that alignment into a phylogenetic tree.
7. Click the run button at the bottom of the page: An intermediate page appears. Wait until your browser displays the Results page.
8. Save your results.

Your results come in three sections:

- Pairwise scores: This is the default output of ClustalW. You can safely ignore the content of this section. It has mostly to do with the pairwise comparisons that ClustalW makes to figure out the guide tree.
- The multiple alignment: This section is in the middle of the output. It contains your alignment. You can display this alignment and save it as a text file if you click the hyperlink that comes just before the alignment. Most ClustalW web servers (including the EBI) output the ClustalW guide tree (a .dnd file). The guide tree is not a phylogenetic tree. To obtain a phylogenetic tree from a ClustalW server, you must cut and paste an actual multiple sequence alignment — not a set of unaligned sequences.

Iterative Methods

A set of methods to produce MSAs while reducing the errors inherent in progressive methods are classified as "iterative" because they work similarly to progressive methods but repeatedly realign the initial sequences as well as adding new sequences to the growing MSA. One reason progressive methods are so strongly dependent on a high-quality initial alignment is the fact that these alignments are always incorporated into the final result, i.e. once a sequence has been aligned into the MSA, its alignment is not considered further. This approximation improves efficiency at the cost of accuracy. By contrast, iterative methods can return to previously calculated pairwise alignments or sub-MSAs incorporating subsets of the query sequence as a means of optimizing a general objective function such as finding a high-quality alignment score.

A variety of subtly different iteration methods have been implemented and made available in software packages; reviews and comparisons have been useful but generally refrain from choosing a "best" technique. The software package PRRN/PRRP uses a hill-climbing algorithm to optimize its MSA alignment score and iteratively corrects both alignment weights and locally divergent or "gappy" regions of the growing MSA. PRRP performs best when refining an alignment previously constructed by a faster method.

Another iterative program, DIALIGN, takes an unusual approach of focusing narrowly on local alignments between subsegments or sequence motifs without introducing a gap penalty. The alignment of individual motifs is then achieved with a matrix representation similar to a dot-matrix plot in a pairwise alignment. An alternative method that uses fast local alignments as anchor points or "seeds" for a slower global-alignment procedure is implemented in the CHAOS/DIALIGN suite.

A third popular iteration-based method called MUSCLE (multiple sequence alignment by log-expectation) improves on progressive methods with a more accurate distance measure to assess the relatedness of two sequences. The distance measure is updated between iteration stages (although, in its original form, MUSCLE contained only 2–3 iterations depending on whether refinement was enabled).

Hidden Markov Models

Hidden Markov models are probabilistic models that can assign likelihoods to all possible combinations of gaps, matches, and mismatches to determine the most likely MSA or set of possible MSAs. HMMs can produce a single highest-scoring output but can also generate a family of possible alignments that can then be evaluated for biological significance. HMMs can produce both global and local alignments. Although HMM-based methods have been developed relatively recently, they offer significant improvements in computational speed, especially for sequences that contain overlapping regions.

Typical HMM-based methods work by representing an MSA as a form of directed acyclic graph known as a partial-order graph, which consists of a series of nodes representing possible entries in the columns of an MSA. In this representation, a column that is absolutely conserved (i.e. that all the sequences in the MSA share a particular character at a particular position) is coded as a single node with as many outgoing connections as there are possible characters in the next column of the alignment. In the terms of a typical hidden Markov model, the observed states are the individual alignment columns and the "hidden" states represent the presumed ancestral sequence from which the sequences in the query set are hypothesized to have descended. An efficient search variant of the dynamic programming method, known as the Viterbi algorithm, is generally used to successively align the growing MSA to the next sequence in the query set to produce a new MSA. This is distinct from progressive alignment methods because the alignment of prior sequences is updated at each new sequence addition. However, like progressive methods, this technique can be influenced by the order in which the sequences in the query set are integrated into the alignment, especially when the sequences are distantly related.

Several software programs are available in which variants of HMM-based methods have been implemented and which are noted for their scalability and efficiency, although properly using an HMM method is more complex than using more common progressive methods. The simplest is POA (partial-order alignment); a similar but more generalized method is implemented in the packages SAM (sequence alignment and modelling system) and HMMER. SAM has been used as a source of alignments for protein structure prediction to participate in the CASP structure prediction experiment and to develop a database of predicted proteins in the yeast species S. cerevisiae. HHsearch is a software package for the detection of remotely related protein sequences based on the pairwise comparison of HMMs. A server running HHsearch (HHpred) was by far the fastest of the 10 best automatic structure prediction servers in the CASP7 and CASP8 structure prediction competitions.

3D PSSM

The 3D PSSM server takes the sequence of protein of interest and attempts to predict its three-dimensional structure and also its function. For each probe, the 3D PSSM library is scanned using the global dynamic programming algorithm. The score for a match between a residue in the probe and a residue in the library sequence is calculated as the sum of the secondary structure, solvation potential and PSSM scores. Three passes of dynamic programming are performed for each query-library sequence match. Each pass differs in the PSSM used for the scoring, with secondary structure and solvation being held constant.

Pass 1: Library sequence is matched to the query PSSM.

Pass 2: Query sequence is matched to the library 1D-PSSM

Pass 3: Query sequence is matched to the library 3D PSSM

Input: The amino acid sequence of protein of interest is submitted in FASTA format with a one line description about the protein. Email ID is indispensible as the results are mailed.

Output: The server generates the results with an in-depth analysis. The browser comes up with proteins that are very similar to the query sequence after searching the nonredundant protein databanks. The secondary structure is predicted by using PSI-BLAST. From the PSI-BLAST results the function of the protein is also predicted. The browser shows e-values which is measure of confidence in the prediction. Based on the matching of the sequences, the server comes up with the list of proteins with similar domains and also classifies the protein into the superfamily based on it's properties using SCOP (structural classification of protein) by scanning the library.

Threading Methods for Protein Structure Prediction

Introduction

As the attempts to sequence entire genomes increases the number of protein sequences by a factor of two each year, the gap between sequence and structural information stored in public databases is growing rapidly. In stark contrast to sequencing techniques, experimental methods for structure determination are time-consuming, and limited in their application, and therefore will not be able to keep pace with the flood of newly characterized gene products. The development of practical methods for predicting protein structure from sequence is therefore of considerable importance in the field of biology. Several different approaches have been used to predict protein structure from sequence, with varying degrees of success. Ab initio methods encompass any means of calculating co-ordinates for a protein sequence from first principles—that is, without reference to existing protein structures. Little success has been seen in this area, with more theory produced than actual useful methodology. Comparative (or homology) modelling, attempts to predict protein structure on the strength of a protein's sequence similarity to another protein of known structure (following the theory that similar sequence implies similar structure). Some success has been achieved, but several limitations to this method, not least of which are its dependence on alignment quality and the existence of a good sequence homologue, indicate it is not applicable to a large fraction of protein sequences. The third main category of protein structure prediction, falling somewhere between comparative modelling and ab initio prediction, is fold recognition, or threading.

Threading Methods

The term 'threading' was first coined in 1992 by Jones et al. The idea behind threading came about from the observation that a large percentage of proteins adopt one of a limited number of folds. In fact, just 10 different folds (the 'superfolds') account for 50% of the known structural similarities between protein superfamilies.

Thus, rather than trying to find the correct structure for a protein from the huge number of all possible conformations available to a polypeptide chain, the correct (or close to correct) structure is likely to have already been observed and already stored in a structural database. Of course, in cases where the target protein shares significant sequence similarity to a protein of known 3D structure, the 'fold recognition' problem is trivial—simple sequence comparison will identify the correct fold. The hope was, however, that threading might be able to detect structural similarities that are not accompanied by any detectable sequence similarity, and this has subsequently been proven to be the case.

Firstly, a library of unique or representative protein structures needs to be derived from the database of all known protein structures.

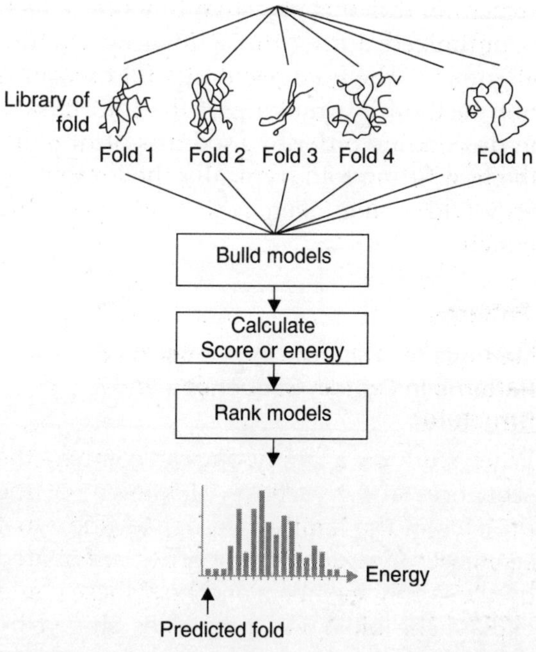

Fig. 4.5: This is an outline of the fold recognition approach to protein structure prediction, and identifies three clear aspects of the problem that need consideration: a fold library, a method for modelling the object sequence on each fold, and a means for assessing the goodness-of-fit between the sequence and the structure

Different groups use different selection criteria for their fold libraries: in some cases, complete protein chains are used in the library, but in other cases, structural domains or even conserved proteins cores are used. Each fold from this library is then considered in turn and the target sequence optimally fitted (or aligned) to each library fold (allowing for relative insertions and deletions in loop regions). Many different algorithms have been proposed for finding this optimal sequence-structure alignment, with most groups using some form of dynamic programming algorithm but other algorithms such as Gibbs sampling or branch-and-bound searching have also been used with some success. Finally, some kind of objective function is needed to determine the goodness of fit between the sequence and the template structure. It is this objective function which is optimized during the sequence-structure alignment. The final result of a fold recognition method is a ranking of the fold library in descending order of goodness of fit, with the best fitting fold (typically the lowest energy fold) being taken as the most probable match.

Patterns

Methods for Discovering Conserved Patterns in Protein Sequences and Structures

When studying a family, one can compare the sequences and structures (if known) of the proteins in the family in order to find what sequence or structure properties are shared by the family members and how these could explain the biological properties shared by the proteins in the family. A description of sequence properties is called a sequence pattern, and a description of structure properties is called a structure pattern. If a pattern is common to a family of proteins, it is called a motif for the family.

An example protein family is the set of proteins containing the classical zinc finger DNA-binding domain. Most of the sequences in this family match the pattern C-x(2,4)-C-x(3)-[LIVMFYWC]-x(8)-H-x(3,5)-H, thus this is a sequence motif for this protein family. A sequence matches this pattern if it contains a C followed by 2–4 arbitrary letters followed by C and 3 arbitrary letters and one of L, I, V, M, F, Y, W, or C, and so on. An example protein family is the set of proteins containing the classical zinc finger DNA-binding domain. Most of the sequences in this family match the pattern C-x(2,4)-C-x(3)-[LIVMFYWC]x(8) -H-x(3,5)-H, thus this is a sequence motif for this protein family. A sequence matches this pattern if it contains a C followed by 2-4 arbitrary letters followed by C and 3 arbitrary letters and one of L, I, V, M, F, Y, W, or C, and so on. This particular pattern describes two cysteines and two histidines that are needed for coordinating the zinc ion in the classical zinc-finger domain, which means that this particular pattern has a direct biological interpretation. Additionally, the pattern can be used for classification, since not only do most sequences in the family match the pattern, but also very few sequences outside the family match the pattern. So, if one finds a match to this pattern in a new sequence, the chances are good that this new protein contains a zinc finger domain and binds to DNA. A large number of patterns have been compiled and collected in different protein family databases.

An example is the PROSITE database which contains more than 1000 protein families and for most of these it gives a pattern which occurs in most of the sequences in the family. The patterns are regular expressions (like the zinc finger pattern given above) or profiles (position specific scoring matrices). Other databases also use local alignments, profiles, and Hidden Markov models (HMMs).

When one knows the structures of some of the proteins in the protein family under study, the structures can be compared and similarities described as patterns or motifs. In the same way that protein structures can be described at different levels (e.g. atom, residue, secondary structure, element level), structure patterns can describe structure properties at different levels. For example, one pattern could describe the packing of four alpha-helices and another pattern could describe the relative position of the cysteines and histidines in the classical zinc finger.

Pattern Descriptions

A very simple type of patterns is substring patterns—a sequence matches a substring pattern if it contains the substring (contiguous word in the sequence). For example, the substring pattern CDEC is matched by all sequences containing CDEC as a substring. This very simple type of patterns can sometimes be useful in the analysis of protein or nucleotide sequences. We will first define the concept of approximate pattern matching and then describe different generalizations of substring patterns.

Exact or approximate matching

When matching a sequence against a substring pattern, one may allow for approximate matching. In this case, a sequence matches the pattern if it contains a substring approximately equal to the pattern. In practice, one defines a measure of distance between a pattern and a substring and sets an upper limit on the distance to be allowed.

One simple way to measure the distance between two strings (or a pattern and a string) is to count the number of character changes needed to transform one into the other. This is called the number of mismatches or Hamming distance and can measure the distance between two strings only if they have equal length. For example, the sequence AGCDFCALKW approximately matches the substring pattern CDEC since the substring CDFC can be transformed into CDED by substituting the F with an E.

More general distance measures allow for insertion and deletion of characters in addition to substitutions. The edit distance between two strings (sequences) is the minimum number of single character insertions/deletions and substitutions needed to transform one into the other. For example, the sequence AGCDDALKW approximately matches the pattern CDED when one allows for an edit distance of more than one, but it does not match if one allows only for one mismatch. When comparing protein sequences, one may also penalize different substitutions differently since some amino acid replacements are found more often in equivalent positions in homologous (evolutionarily related) proteins. For example substitutions can be penalized using a substitution matrix, e.g. PAM-matrices or BLOSUM matrices .

In order to find the edit distance between two sequences, one can use the dynamic programming algorithm which can also be used when substitution matrices are used. For substring patterns, the matching problem is very similar to local pairwise alignment and database searching where speed-ups can be used, e.g. BLAST, FASTA.

PROSITE Patterns

Using substring patterns and approximate matching one cannot specify that some pattern positions are compulsory (for instance, that there must be cysteines or histidines for binding zinc) while other positions are allowed to vary more freely. Figure 4.6 illustrates this by showing four segments (substrings) containing the zinc finger motif and a consensus sequence (substring pattern) for the four. If this substring pattern is to match the four segments shown, one needs to allow up to 17 mismatches.

Local sequence alignment	Distance to consensus
EKPFACDFCGRKFARSDERKRHTKIHLRQKE	17
KHPFQCAICMRNFSRSDHLTTHIRTHTGEKP	1
HKPFQCRICMRNFSRSDGLTTGIRTGTGEKP	0
HKPFQCRICMRNFSRSDGLTTHIRTHTGEKP	0
GKPFQCRICMRNFSRSDHLTTHIRTHTGEKP	
Consensus sequence	

Fig. 4.6: Example of a local alignment of sequence segments containing the classical zinc finger motif. The consensus sequence below shows for each position which amino acid is the most frequent, and to the right it is shown for each segment the number of mismatches between the segment and the consensus

A better way would be to allow no mismatches in the conserved positions, including the cysteines and histidines, and allow the remaining positions to be filled by any amino acid.

Some pattern languages allow a description of which amino acids are allowed at each position. For instance, in the PROSITE database, one uses a pattern language which is a subtype of regular expressions. Each pattern consists of a sequence of pattern elements that are of the following types:

1. Single residue: matches one letter identical to the residue in the sequence, e.g. R matches an R in the sequence.
2. Set of residues given in square brackets: matches any one sequence letter contained in the set, e.g. [KER] matches any one of K, E, or R in the sequence.
3. Set of residues given in curly brackets: matches any one sequence letter not in the set, e.g. {KER} matches any letter except K, E, and R in the sequence.
4. Wildcard x: matches any one letter in a sequence additionally:
   a. Single pattern elements can be followed by parentheses (i, j) which means that the sequence can contain between i and j (inclusive) letters each matching the preceding pattern element. For example, x(3,5) matches between 3 and 5 arbitrary sequence letters.
   b. The pattern can start with '<' meaning that the pattern should match from the beginning of a sequence.
   c. The pattern can end with '>' meaning that the pattern should match until the end of a sequence.
   d. The pattern elements are separated by hyphens '-'. Consecutive pattern elements should be matched by consecutive sequence symbols, so for example the pattern C-x(2,3)-[DE] matches any sequence containing a C followed by two or three arbitrary letters followed by a D or an E.

Pattern Databases

A number of different databases for storing information about protein families and motifs have been established during the last 10 years. They differ in a number of ways. Firstly, they differ in how they represent the patterns for each family. Secondly, some of them are constructed manually (both sequence grouping and pattern definition) while others are made (to a varying degree) automatically.

PROSITE: One of the most widely used databases is PROSITE , in which, or each family, one or several patterns and/or profiles are given in the format described above.

Profiles are described using the generalised profile syntax. Statistics are given which describe the patterns' ability to discriminate between family members and other sequences given in the SWISS-PROT protein sequence database, in the form of the number of false positives and false negatives. Also, a number of unknowns is given which is the number of sequences in SWISS-PROT which match the pattern, but for which it is not yet known whether it belongs to the family or not.

Table 4.3: Summary information about a few family databases

Database	*Pattern type*	*URL*
PROSITE	Reg.exp and profiles	http://expasy.hcuge.ch/sprot/prosite.html
BLOCKS	Blocks (ungapped local alignments)	http://www.blocks.fhcrc.org/
Prints	Blocks	http://www.biochem.ucl.ac.uk/bsm/dbbrowser/PRINTS/
Identify	Reg.exp	http://motif.stanford.edu/identify
Pfam	HMMs	http://www.sanger.ac.uk/pfam/

Fig. 4.7 shows a PROSITE entry giving a signature pattern (motif) for the actinin-type actin-binding domain. We will explain the most important (for our purpose) parts of the entry. First, the ID and AC lines give, respectively, the name and the accession number of the PROSITE entry. The pattern is given on the PA line (the pattern can continue over several PA lines, then end of the pattern being marked by period sign). The NR lines give statistics about the pattern's discriminatory power with respect to the SWISS-PROT


```
ID   ACTININ_1; PATTERN.
AC   PS00019;
DT   APR-1990 (CREATED); NOV-1997 (DATA UPDATE); NOV-1997 (INFO UPDATE).
DE   Actinin-type actin-binding domain signature 1.
PA   [EQ]-x(2)-[ATV]-[FY]-x(2)-W-x-N.
NR   /RELEASE=35,69113;
NR   /TOTAL=55(46); /POSITIVE=35(28); /UNKNOWN=0(0); /FALSE_POS=20(18);
NR   /FALSE_NEG=0; /PARTIAL=1;
CC   /TAXO-RANGE=??E??; /MAX-REPEAT=2;
DR   P12814, AAC1_HUMAN, T; P35609, AAC2_HUMAN, T; Q08043, AAC3_HUMAN, T;
DR   Q99001, AACB_CHICK, T; Q90734, AACN_CHICK, T; P20111, AACS_CHICK, T;
DR   P05094, AACT_CHICK, T; P05095, AACT_DICDI, T; P18091, AACT_DROME, T;
DR   P21333, ABP2_HUMAN, T; P11533, DMD_CHICK , T; P11532, DMD_HUMAN , T;
DR   P11531, DMD_MOUSE , T; P19179, FIMB_CHICK, T; P54680, FIMB_DICDI, T;
DR   P32599, FIMB_YEAST, T; P13466, GELA_DICDI, T; Q14651, PLSI_HUMAN, T;
DR   P13796, PLSL_HUMAN, T; Q61233, PLSL_MOUSE, T; P13797, PLST_HUMAN, T;
DR   Q63598, PLST_RAT  , T; Q00963, SPCB_DROME, T; P11277, SPCB_HUMAN, T;
DR   P15508, SPCB_MOUSE, T; Q01082, SPCO_HUMAN, T; Q62261, SPCO_MOUSE, T;
DR   P46939, UTRO_HUMAN, T;
DR   P11530, DMD_RAT   , P;
DR   P13688, BGP1_HUMAN, F; P06731, CCEM_HUMAN, F; P31997, CGM6_HUMAN, F;
DR   P40782, CYP1_CYNCA, F; P10474, GUNB_CALSA, F; P40199, NCA_HUMAN , F;
DR   P11462, PBG1_HUMAN, F; P11463, PBGC_HUMAN, F; P11464, PBGD_HUMAN, F;
DR   Q00887, PSGB_HUMAN, F; P35853, PUR1_LACCA, F; P14410, SUIS_HUMAN, F;
DR   P23739, SUIS_RAT  , F; P28668, SYEP_DROME, F; P07814, SYEP_HUMAN, F;
DR   P42954, TAGH_BACSU, F; P09301, UL07_VZVD , F; P52583, VGR2_COTJA, F;
3D   1KSR; 1DRO; 1BTN; 1MPH;
DO   PDOC00019;
```

Fig. 4.7: Example of a PROSITE entry

database. The first NR line says that the statistics are with respect to release 35 of SWISS-PROT, which has 69113 sequence entries. The next NR line gives the number of matches of different categories (true positives, false positives, false negatives, etc.) in the SWISS-PROT database. Each is given as x(y) meaning that there are x matches to the pattern in y different sequences.

The DR lines give references to the corresponding SWISS-PROT entries both by their names and accession numbers. Each reference is on the. form 'AC, ID, Status;' where Status is one of T (true positive), N (false negative), F (false positive), P (partial), and ? (unknown). Finally, the 3D line gives names of PDB (Protein Data Bank) entries containing structures of proteins in the family, and the DO line gives the accession number of the entry in the PROSITE documentation part corresponding to this entry.

Each entry in the PROSITE documentation part gives a description of the family and explains the biological significance of the signature patterns. It also gives literature references and one or several experts that can be contacted for more information about the family. The PROSITE database is largely maintained manually; new families, profiles and patterns are carefully scrutinized.

BLOCKS: Another protein family database is BLOCKS which contains the same families as PROSITE, but instead of giving patterns or profiles, it gives a set of blocks for each family. A block is an ungapped local multiple alignment. The database is constructed fully automatically. For each family, the member sequences (as given in PROSITE) are subjected to pattern discovery and local alignment methods. The blocks can also be linked in chains. It is recommended that a query sequence is matched against both PROSITE and BLOCKS, because even though they describe the same families, the patterns and the blocks in a sense have complementary strengths when used as classifiers.

PRINTS: The PRINTS database is a collection of fingerprints and is constructed semi-automatically. A fingerprint is defined as a list of motifs, each motif being a local ungapped alignment. The fingerprints are made by first manually making an alignment of some family members and then iteratively scanning a database of protein sequences, adding new members, updating the fingerprint, scanning again until convergence (no new family members are found). Each entry in PRINTS, gives the local alignments corresponding to the motif and information about partial matches, etc. On the website of PRINTS, a tool Finger-PRINTScan, is available for scanning a query sequence against the database. The output can be visualized showing the position of the motif matches in the query sequence.

Pfam: Pfam is a database of protein families and corresponding profile-HMMs. Pfam uses the HMMER2 package to provide tools for making the HMMs in the 'first place and then for searching them. A search against Pfam will provide you with three ways of deciding confidence in the matches. The first is a classical e-value (expectation value) which generally is considered significant if it is below 1.0 for individual searches. The second is the Bits score which is derived from the underlying scoring scheme used to score the match between the sequence and the profile-HMMs.

It is related to the Bayesian inference of the probability of the match. A final check is provided by a manually derived cut-off which an 'expert' has chosen to separate the true examples from false examples. These cut-offs are chosen conservatively so that, to the researcher's knowledge, they do not misclassify any protein. This can mean that, in some cases, known trues are missed using this cut-off.

Identify: Identify is a database of patterns of the same form as used in PROSITE but without flexible length wildcards. It contains patterns for the families in the BLOCKS and PRINTS databases. The patterns were constructed from the ungapped alignments (blocks) in the BLOCKS and PRINTS databases by using a pattern finding program (EMOTIF). For each block, there can be several patterns so that each pattern matches a subset of the sequences.

Structure Motifs

Finding recurring patterns (motifs) in protein structures help to better understand the rules underlying the formation of protein structures. Since structure is better conserved during evolution than sequence, structural similarities can also help to identify remote evolutionary relationships. Structure motifs can help in approaching the structure prediction problem and in assigning function to proteins. Structure motifs can represent common structural features at different levels. For example, they can represent packing of secondary structure elements, local packing of residues, and atom coordinates of binding atoms in active site (or ligand binding) residues.

In order to find recurring patterns in protein structures, one can use methods for the comparison of protein structures. A number of such methods have been developed, most of them for comparing pairs of structures, but also some for multiple structure comparison. The methods differ in what similarities they are able to find. Some represent the structures as composed of secondary structure elements (alpha helices, beta strands, and loops) and have provided methods for finding patterns of conserved patterns at this level. Other methods find patterns of residues (or atoms) that have similar configurations in space. An important difference between methods is whether they require matched elements to be in the same order along the proteins' primary structure.

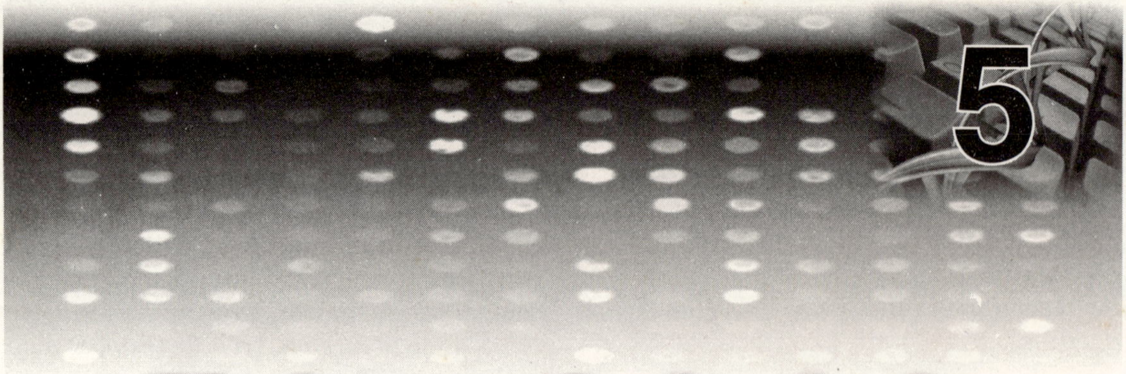

5 Phylogenetic Analysis

Phylogenetic analysis of nucleic acid and protein sequences is presently and will continue to be an important area of sequence analysis. It is determination of how the family might have been derived during evolution. Evolutionary relationship among the sequences is depicted with the help of phylogenetic trees. Similarities and divergence among related biological sequences revealed by sequence alignment often have to be rationalized and visualized in the context of phylogenetic trees. The evolutionary relationships among the sequences are depicted by placing the sequences as outer branches on a tree. The branching relationships on the inner part of the tree then reflect the degree to which different sequences are related. Two sequences that are very much alike will be located as neighboring outside branches and will be joined to a common branch beneath them. The object of phylogenetic analysis is to discover all of the branching relationships in the tree and the branch lengths.

Phylogenetics is a special kind of phylogeny that relies on the comparison of equivalent genes coming from several species for reconstructing the genealogic tree of these species and finding out who is the closest relative of whom in the family. If necessary, you can also apply phylogenetic methods to the various genes of a gene family to reconstruct the history of the gene family by the same means.

Phylogenetics can be studied in three ways. In phenetics, species are grouped with others they resemble phenotypically and all the characters are taken into account. In Cladistics, species are grouped only with those that share derived characters, that is characters that were not present in their distant ancestors. The third approach evolutionary systematics, incorporates both phonetic and cladistic principles. Cladistics is accepted as the best method available for phylogenetic analysis because it accepts and employs current evolutionary theory, i.e. that speciation occurs by bifurcation (cladogenesis).

Major Reasons Why We may Want to Use Phylogenetic Analysis

- Determining the closest relatives of the organism that you are interested in: For instance, if you are studying a new bacterium, you can sequence its ribosomal RNA and place it on a phylogenetic tree computed with all known ribosomal RNAs. This can give you a fairly good idea of who this bacterium really is.
- Discovering the function of a gene: If you are studying a gene, you can use phylogenetic trees to be sure that the gene you

are interested in is orthologous (more about that in a minute) to another well-characterized gene in another species.

- Retracing the origin of a gene: Most genes within a genome travel together through evolutionary time. However, from time to time, individual genes may jump from one species to another — for instance, piggybacking a virus infection. Phylogenetic trees are a great way to reveal such events, which are called horizontal (or lateral) transfers.

Two genes are orthologous if they come from two different organisms, derive from a common ancestor gene, and have been separated only by speciation events (as opposed to gene duplications). In theory, two genes that are orthologous often have the same exact function (have similar roles) in the two different organisms they come from. On the other hand, when the two genes arose from gene duplication, we say that they are paralogous. Two genes that are paralogous belong to the same family but are more likely to have different functions than they would if they were orthologous.

For example, if joystick of an aeroplane were a gene, it would be orthologous with the steering wheel of a car. In contrast, the left front wheel of a car is paralogous with the right front wheel.

A typical phylogenetic tree is as shown in the Fig. 5.1 .

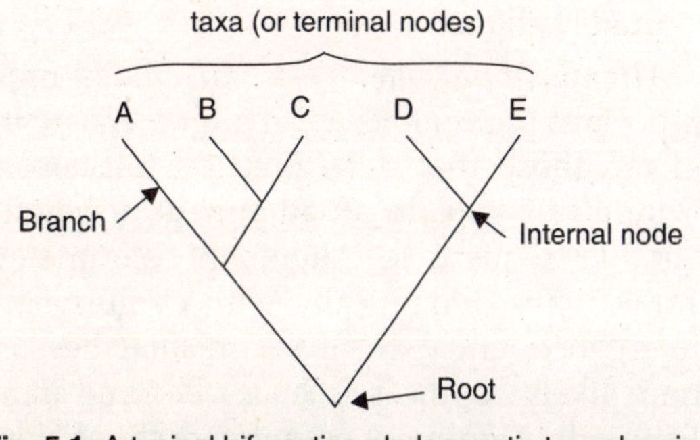

Fig. 5.1: A typical bifurcating phylogenetic tree showing root, internal nodes, terminal nodes and branches

The line in the tree are called as branches. At the tips of the branches are species or sequences known as taxa (the singular form is taxon) or operational taxonomic units. The connecting point where two adjacent branches join is called a node, which represents an inferred ancestor of extant taxa. The bifurcating point at the very bottom of the tree is the root node, which represents the common ancestor of all members of the tree.

Forms of Tree Representation

In a phylogram, the branch lengths represent the amount of evolutionary divergence. Such trees are said to be scaled. The scaled trees have the advantage of showing both the evolutionary relationships and information about the relative divergence time of the branches. In a cladogram, however, the external taxa line up neatly in a row or column. Their branch lengths are not proportional to the number of evolutionary changes and thus have no phylogenetic meaning. In such an unscaled tree, only the topology of the tree matters, which shows the relative ordering of the taxa (Fig. 5.2).

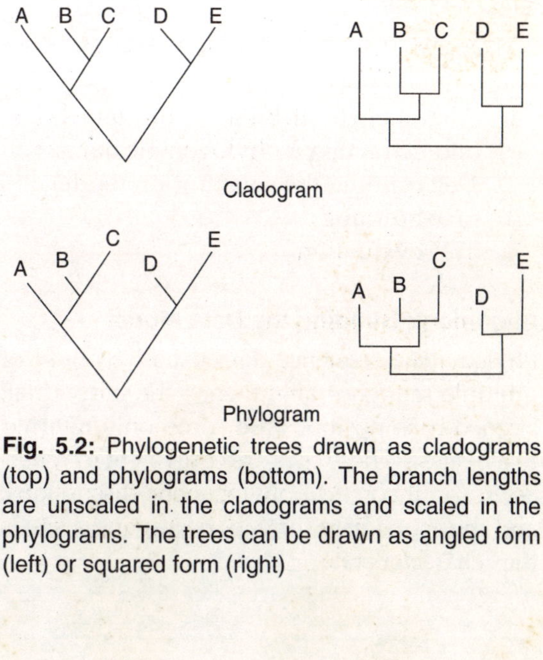

Fig. 5.2: Phylogenetic trees drawn as cladograms (top) and phylograms (bottom). The branch lengths are unscaled in the cladograms and scaled in the phylograms. The trees can be drawn as angled form (left) or squared form (right)

Phylogenetic tree can be either rooted or unrooted. An unrooted phylogenetic tree does not assume knowledge of a common ancestor, but only positions the taxa to show their relative relationships. Because there is no indication of which node represents an ancestor, there is no direction of an evolutionary path in an unrooted tree. To define the direction of an evolution path, a tree must be rooted. In a rooted tree, all the sequences under study have a common ancestor or root node from which a unique evolutionary path leads to all other nodes. Obviously, a rooted tree is more informative than an unrooted one.

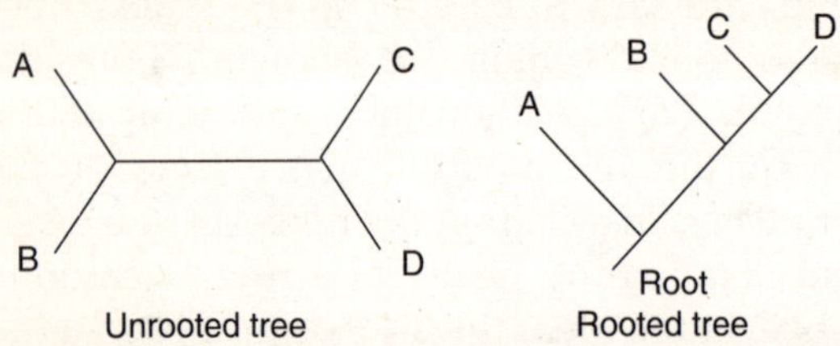

Fig. 5.3: An illustration of rooted versus unrooted trees. A phylogenetic tree without definition of a root is unrooted (left). The tree with a root is rooted (right)

Phylogenetic Data Analysis: The Four Steps

A straightforward phylogenetic analysis consists of four steps:

1. Alignment (both building the data model and extracting a phylogenetic dataset)
2. Determining the substitution model
3. Tree building
4. Tree evaluation.

Alignment: Building the Data Model

Phylogenetic sequence data usually consist of multiple sequence alignments; the individual, aligned-base positions are commonly referred to as "sites." These sites are equivalent to "characters" in theoretical phylogenetic discussions, and the actual base (or gap) occupying a site is the "character state."

Steps in building the alignment include selection of the alignment procedure(s) and extraction of a phylogenetic data set from the alignment. The latter procedure requires determination of how ambiguously aligned regions and insertion/deletions (referred to as indels, or gaps) will be treated in the tree-building procedure. A typical alignment procedure involves the application of a program such as CLUSTAL W, followed by manual alignment editing and submission to a tree-building program.

Phylogenetic criteria preferred: Some computational multiple alignment methods align sequences strictly based on the order they receive them (the input order) without any consideration of their relationship. However, many current methods (e.g. CLUSTAL W, PileUp, ALIGN in ProPack) align according to an explicitly phylogenetic criterion (a "guide tree"). These guide trees are generated on the basis of initial pairwise sequence alignments. SAM and MACAW are examples of multiple alignment programs that do not explicitly invoke phylogenetic criteria, although it is possible to manipulate parameters in these programs to mimic phylogenetic processes.

Theory holds that more closely related sequences should be aligned first and then the resulting groups of sequences, which may be less related to one another but still have a common ancestor, should share the same ancestral indels. This means that they could then be more accurately aligned.

Alignment parameter estimation: The most important parameters in an alignment method are those that determine the placement of indels or gaps in an alignment of length-variable sequences. Alignment parameters should vary dynamically with evolutionary divergence, such that base mismatches are more likely as the sequences become more divergent. Alignment parameters should also be adjusted to prevent closely related, over-represented sequences from adversely influencing the alignment of underrepresented sequences. This is accomplished by down-weighting the alignment score contribution of closely related sequences. These dynamic parameter adjustments are both implemented in CLUSTAL W, whereas sequence weighting is implemented in SAM.

Which Alignment Procedure is Best for Phylogenetic Analysis?

The short answer is "the method that is closest to understanding the evolutionary relationships between the sequences being examined." Unless the actual phylogenetic relationships are known beforehand, there is no clear way to determine which alignment procedure is best for a given phylogenetic analysis. In general, it is inadvisable to simply subject a computer-generated alignment to a tree-building procedure because the latter is blind to errors in the former. This caution especially applies to tree-building programs included in alignment packages (e.g. CLUSTALW and TREE in ProPack) because the tree-building methods in these programs are not rigorous. However, as long as the entire alignment is scrutinized in view of independent phylogenetic evidence, methods such as CLUSTAL W that utilize some degree of phylogenetic criteria are among the best currently available.

Following points should be considered when constructing a multiple sequence alignment for a phylogenetic analysis:

The alignment step in phylogenetic analysis is one of the most important because it produces the data set on which models of evolution are used.

- It is not uncommon to edit the alignment, deleting unambiguously aligned regions and inserting or deleting gaps to more accurately reflect probable evolutionary processes that led to the divergence between sequences.
- It is useful to perform phylogenetic analyses based on a series of slightly modified alignments to determine how ambiguous regions in the alignment affect the results and what aspects of the results one may have more or less confidence in.

Determining the Substitution Model

The substitution model should be given the same emphasis as alignment and tree-building. Substitution model influences both alignment and tree-building; hence, a recursive approach is warranted. At the present time, two elements of the substitution model can be computationally assessed for nucleotide data but not for amino acid or codon data. One element is the model of substitution between particular bases; the other is the relative rate of overall substitution among different sites in the sequence. Simple computational procedures have not been developed for assessing more complex variables (e.g. site- or lineage specific substitution models).

Phylogenetic Tree Construction Methods

There are currently two main categories of tree-building methods, each having advantages and limitations. The first category is based on discrete characters, which are molecular sequences from individual taxa. The basic assumption is that characters at corresponding positions in a multiple sequence alignment are homologous among the sequences involved. Therefore, the character states of the common ancestor can be traced from this dataset. Another assumption is that each character evolves independently and is therefore treated as an individual evolutionary unit. The second category of phylogenetic methods is based on distance, which is the amount of dissimilarity between pairs of sequences, computed on the basis of sequence alignment. The distance-based methods assume that all sequences involved are homologous and that tree branches are additive, meaning that the distance between two taxa equals the sum of all branch lengths connecting them. More details on procedures and assumptions for each type of phylogenetic method are described.

Distance-matrix methods

Distance-matrix methods of phylogenetic analysis explicitly rely on a measure of "genetic distance" between the sequences being classified, and therefore they require an MSA as an input. Distance is often defined as the fraction of mismatches at aligned positions, with gaps either ignored or counted as mismatches. Distance methods attempt to construct an all-to-all matrix from the sequence query set describing the distance between each sequence pair. From this is constructed a phylogenetic tree that places closely related sequences under the same interior node and whose branch lengths closely reproduce the observed distances between sequences. Distance-matrix methods may produce either rooted or unrooted trees, depending on the algorithm used to calculate them. They are frequently used as the basis for progressive and iterative types of multiple sequence alignments. The main disadvantage of distance-matrix methods is their inability to efficiently use information about local high-variation regions that appear across multiple subtrees.

Neighbor-joining: Neighbor-joining methods apply general data clustering techniques to sequence analysis using genetic distance as a clustering metric. The simple neighbor-joining method produces unrooted trees, but it does not assume a constant rate of evolution (i.e. a molecular clock) across lineages. Its relative, UPGMA (unweighted pair group method with Arithmetic mean) produces rooted trees and requires a constant-rate assumption, i.e. it assumes an ultrametric tree in which the distances from the root to every branch tip are equal.

Fitch-Margoliash method: The Fitch-Margoliash method uses a weighted least squares method for clustering based on genetic distance. Closely related sequences are given more weight in the tree construction process to correct for the increased inaccuracy in measuring distances between distantly related sequences. The distances used as input to the algorithm must be normalized to prevent large artifacts in computing relationships between closely related and distantly related groups.

The distances calculated by this method must be linear; the linearity criterion for distances requires that the expected values of the branch lengths for two individual branches must equal the expected value of the sum of the two branch distances—a property that applies to biological sequences only when they have been corrected for the possibility of back mutations at individual sites. This correction is done through the use of a substitution matrix such as that derived from the Jukes-Cantor model of DNA evolution. The distance correction is only necessary in practice when the evolution rates differ among branches.

The least-squares criterion applied to these distances is more accurate but less efficient than the neighbor-joining methods. An additional improvement that corrects for correlations between distances that arise from many closely related sequences in the data set can also be applied at increased computational cost. Finding the optimal least-squares tree with any correction factor is NP-complete, so heuristic search methods like those used in maximum-parsimony analysis are applied to the search through tree space.

Minimum evolution: Minimum evolution (ME) constructs a tree with a similar procedure, but uses a different optimality criterion that finds a tree among all possible trees with a minimum overall branch length. The optimality criterion relies on the formula:

$$S = \sum bi$$

where *bi* is the ith branch length. Searching for the minimum total branch length is an indirect approach to achieving the best fit of the branch lengths with the original dataset. Analysis has shown that minimum evolution in fact slightly outperforms the least square-based FM method.

Unweighted pair group method using arithmetic average: The simplest clustering method is UPGMA, which builds a tree by a sequential clustering method. Given a distance matrix, it starts by grouping two taxa with the smallest pairwise distance in the distance matrix. A node is placed at the midpoint or half distance between them. It then creates a reduced matrix by treating the new cluster as a single taxon.

The distances between this new composite taxon and all remaining taxa are calculated to create a reduced matrix. The same grouping process is repeated and another newly reduced matrix is created. The iteration continues until all taxa are placed on the tree. The last taxon added is considered the outgroup producing a rooted tree.

The basic assumption of the UPGMA method is that all taxa evolve at a constant rate and that they are equally distant from the root, implying that a molecular clock is in effect. However, real data rarely meet this assumption. Thus, UPGMA often produces erroneous tree topologies. However, owing to its fast speed of calculation, it has found extensive usage in clustering analysis of DNA microarray data.

Character-based methods

The character-based methods have little in common with each other, besides the use of the character data at all steps in the analysis. This allows the assessment of the reliability of each base position in an alignment on the basis of all other base positions.

Maximum parsimony (MP): Maximum parsimony is a method of identifying the potential phylogenetic tree that requires the smallest total number of evolutionary events to explain the observed sequence data. Some ways of scoring trees also include a "cost" associated with particular types of evolutionary events and attempt to locate the tree with the smallest total cost. This is a useful approach in cases where not every possible type of event is equally likely, e.g. when particular nucleotides or amino acids are known to be more mutable than others.

The most naive way of identifying the most parsimonious tree is simple enumeration—considering each possible tree in succession and searching for the tree with the smallest score. However, this is only possible for a relatively small number of sequences or species because the problem of identifying the most parsimonious tree is known to be NP-hard; consequently a number of heuristic search methods for optimization have been developed to locate a highly parsimonious tree, if not the most optimal in the set. Most such methods involve a steepest descent-style minimization mechanism operating on a tree rearrangement criterion.

Maximum likelihood (ML): ML turns the phylogenetic problem inside out. ML searches for the evolutionary model, including the tree itself, that has the highest likelihood of producing the observed data. In practice, ML is derived for each base position in an alignment.

The likelihood is calculated in terms of the probability that the pattern of variation at a site would be produced by a particular substitution process, given a particular tree and the overall observed base frequencies. The likelihood becomes the sum of the probabilities of each possible reconstruction of substitutions under a particular substitution process. The likelihoods for all the sites are multiplied to give an overall "likelihood of the tree" (i.e. the probability of the data given the tree and the substitution process). As one can imagine, for one particular tree, the likelihood of the data is low at some sites and high at others. For a "good" tree, many sites will have higher likelihood, so the product of likelihoods is high. For a "poor" tree, the reverse will be true.

The substitution model should be optimized to fit the observed data. For example, if there is a transition bias, evident by an inordinate number of sites that include only purines or pyrimidines, the likelihood of the data under a model that assumes no bias will never be as good as one that does.

Likewise, if a substantial proportion of the sites are occupied by a single base and another substantial proportion have equal base frequencies, the likelihood of the data under a model that assumes that all sites evolve equally will be less than that of a model that allows rate heterogeneity. Modifying the substitution parameters, however, modifies the likelihood of the data associated with particular trees. Thus, the tree yielding the highest likelihood under one substitution model might yield much lower likelihood under another.

Phylogenetic Tree Evaluation

After phylogenetic tree construction, the next step is to statistically evaluate the reliability of the inferred phylogeny. There are two questions that need to be addressed. One is how reliable the tree or a portion of the tree is; and the second is whether this tree is significantly better than another tree. To answer the first question, we need to use analytical resampling strategies such as boot-strapping and jack-knifing, which repeatedly resample data from the original dataset. For the second question, conventional statistical tests are needed.

Boot-strapping: Boot-strapping is a way of testing the reliability of the dataset. It is the creation of pseudoreplicate datasets by resampling. Boot-strapping allows you to assess whether the distribution of characters has been influenced by stochastic effects. In phylogenetic analyses, nonparametric boot-strapping is the most commonly used method. The pseudoreplicate datasets are generated by randomly sampling the original character matrix to create new matrices of the same size as the original. The frequency with which a given branch is found is recorded as the bootstrap proportion. These proportions can be used as a measure of the reliability (within limitations) of individual branches in the optimal tree.

Thus bootstrap analysis:

- Is a statistical method for obtaining an estimate of error
- Is used to evaluate the reliability of a tree
- Is used to examine how often a particular cluster in a tree appears when nucleotides or amino acids are resampled.

However, if the original dataset is biased, a cluster may be regarded as statistically significant, even if it is a wrong one!

How is boot-strapping and the construction of a consensus tree carried out in practice?

Take a dataset consisting of in total n sequences with m sites each. A number of resampled datasets of the same size (n m) as the original dataset is produced. However, each site is sampled at random and no more sites are sampled than there were original sites. In order to be statistically significant, the number of the datasets should be high and equal or higher than the number of individual sites present in the dataset. Our example dataset consists of in total 4 sequences with 10 sites each. When three new datasets are prepared by random sampling of sites, the following three sample sets of data can be obtained:

```
Sample 1                                 <- number of times each site is sampled
    0 1 2 0 3 0 1 2 0 1
_____________________

A   A G G C U C C A A A              A   G G G U U U C A A A
B   A G G U U C G A A A              B   G G G U U U G A A A
C   A G C C C C G A A A              C   G C C C C C G A A A
D   A U U U C C G A A C              D   U U U C C C G A A C
```

	A	B	C
B	1		
C	6	5	
D	8	7	4

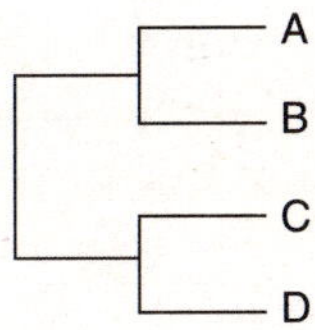


```
Sample 2
   1 0 0 0 2 2 2 0 0 3
_____________________

A   A G G C U C C A A A              A   A U U C C C C A A A
B   A G G U U C G A A A              B   A U U C C G G A A A
C   A G C C C C G A A A              C   A C C C C G G A A A
D   A U U U C C G A A C              D   A C C C C G G C C C
```

	A	B	C
B	2		
C	4	2	
D	7	5	3

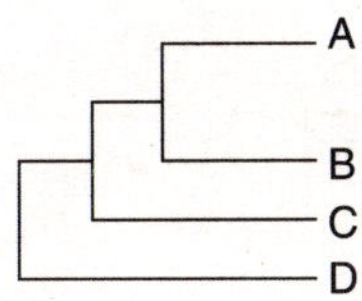


```
Sample 3
   1 0 0 0 2 2 2 0 0 3
_____________________

A A G G C U C C A A A              A   A U U C C C C A A A
B A G G U U C G A A A              B   A U U C C G G A A A
C A G C C C C G A A A              C   A C C C C G G A A A
D A U U U C C G A A C              D   A C C C C G G C C C
```

	A	B	C
B	1		
C	3	2	
D	6	3	4

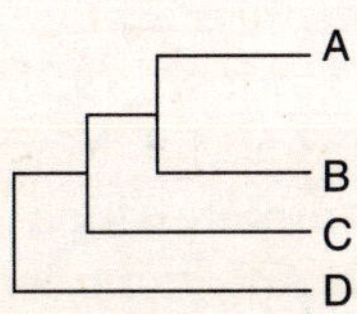

5

A large number of datasets (between hundred and thousand, depending on computer power) and the same number of different trees are so generated. In this specific case, taxa A and B form a cluster in all three trees, while C clusters with D in only one tree. There exist specialized programs, such as the program Consense in the Phylip package of Joe Felsenstein, that are able to analyze all the resulting trees and prepare the most likely tree or consensus tree from those data.

The resulting consensus tree for our small dataset is shown below. The number of times each branch point or node occurred (the so-called bootstrap proportion) is indicated at each node.

Result

```
A   A G G C U C C A A A
B   A G G U U C G A A A
C   A G C C C C G A A A
D   A U U U C C G A A C
```

	A	B	C
B	2		
C	3	3	
D	6	4	4

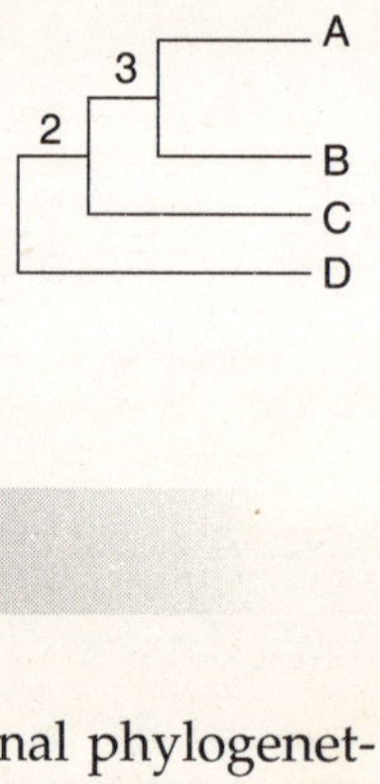

Tools For Carrying Out Phylogenetic Analysis

PHYLIP

PHYLIP is a free computational phylogenetics package of programs for inferring evolutionary trees (phylogenies). The name is an acronym for PHYLogeny Inference Package. It consists of 35 portable programs, i.e. the source code is written in C and precompiled executables are available for Windows (95/98/NT/2000/me/XP), Mac OS 8 and 9, Mac OS X, and Linux systems. Complete documentation is written for all the programs in the package and is part of the package. The author of this package is Joseph Felsenstein, Professor in the Department of Genome Sciences and the Department of Biology at the University of Washington, Seattle.

Methods (implemented by each program) that are available in the package include parsimony, distance matrix, and likelihood methods, including boot-strapping and consensus trees. Data types that can be handled include molecular sequences, gene frequencies, restriction sites and fragments, distance matrices, and discrete characters.

Each program is controlled through a menu, which asks the users which options they want to set, and allows them to start the computation. The data is read into the program from a text file, which the user can prepare using any word processor or text editor (but it is important that this text file not be in the special format of that word processor — it should instead be in flat ASCII or text only format). Some sequence analysis programs such as the Clustal W alignment program can write data files in the PHYLIP format. Most of the programs look for the data in a file called infile -- if they do not find this file they then ask the user to type in the file name of the data file.

Output is written on to files with names like outfile and outtree. Trees written on to outtree are in the Newick format, an informal standard agreed to in 1986 by authors of a number of major phylogeny packages.

The program package is downloadable from http://evolution.genetics.washington.edu/phylip.html.

An online version is also available at http://bioweb.pasteur.fr/seqanal/phylogeny/phylip-uk.html.

PAUP

PAUP (phylogenetic analysis using parsimony and other methods) is a program for phylogenetic analysis using parsimony, maximum likelihood, and distance methods. The program features an extensive selection of analysis options and model choices, and accommodates DNA, RNA, protein and general data types. Among the many strengths of the program are the rich array of options for dealing with phylogenetic trees including importing, combining, comparing, constraining, rooting and testing hypotheses. PAUP version 4.0 is a major upgrade and new release of the software package for inference of evolutionary trees, for use in Macintosh, Windows, UNIX/VMS, or DOS-based formats.

PAUP performs the nonparametric bootstrap for distance, MP, and ML, using all options available for tree-building with these methods. When a bootstrap or jack-knife with MP is under way, MAXTREES should be set between 10 and no more than 100. This is because poorly resolvable portions of an MP tree will usually be even less resolvable with resampled data; hence, a replicate could find astronomical numbers of equal-length trees. Because tree branches weakly supported by the full data set will not have high bootstrap or jack-knife values, limiting MAXTREES will have little, if any, bearing on the results, especially if the number of replicates is increased to, say, 1,000. In addition, PAUP performs the Kishino-Hasegawa test to compare MP or ML trees, computes four types of consensus of multiple trees (usually used for multiple equal-length MP trees), computes stepwise differences between MP trees, and evaluates signal conflicts between specified partitions of sites (e.g. nuclear and organellar sequence data in a combined analysis).

TREE-PUZZLE

It is a program performing quartet puzzling. The advantage is that it allows various substitution models for likelihood score estimation and incorporates a discrete γ model for rate heterogeneity among sites. Because of the heuristic nature of the program, it allows ML analyses of large datasets. The resulting puzzle trees are automatically assigned puzzle support values to internal branches.

5

These values are percentages of consistent quartet trees and do not have the same meaning as bootstrap values. TREE-PUZZLE version 5.0 is available for Mac, UNIX, and Windows and can be downloaded from www.tree-puzzle.de/. There is also an online version of the program available at: http://bioweb.pasteur.fr/seqanal/interfaces/Puzzle.html.

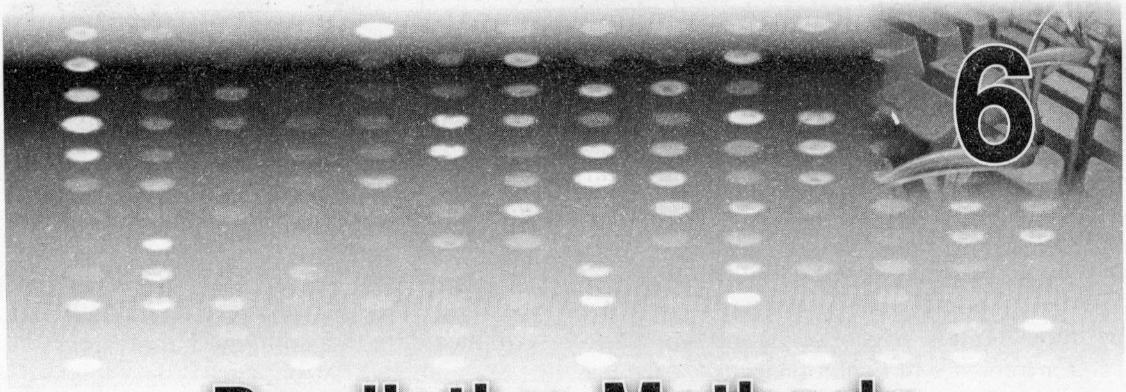

6

Predictive Methods

Approaches to Gene Identification

There are two approaches in the development of sequence analysis techniques for gene identification.

Template methods attempt to compose more or less concise descriptions of prototype objects and then identify genes by matching to such prototypes. The method identifies statistical patterns that distinguish coding from noncoding DNA sequences. A good example is the use of consensus sequences in identifying promoter elements or splice sites. Curation of data is a prerequisite to develop pattern recognition algorithms for identifying features of biological interest for which well-cleansed, nonredundant databases are needed. Information on nonredundant, gene-oriented clusters can be found at UniGene (http://www.ncbi.nlm.nih.gov/UniGene/) and TIGR Gene Indices (http://www.tigr.org/tdb/tdb.html).

Look-up methods, on the other hand, attempt to identify a gene or gene component by finding a similar known object in available databases. The method relies on finding coding regions by their homology to known expressed sequences. An excellent example is the search for genes by trying to find a similarity between the query sequence and the contents of the sequence databases. Some representative human gene expression databases include Cellular Response Database (http://LH15.umbc.edu/crd), GeneCards (http://bioinformatics.weizmann.ac.il/cards/), Globin Gene Server (http://globin.csc.psu.edu), Human Developmental Anatomy (http://www.ana.ed.ac.uk/anatomy/database/), and Merck Gene Index (http://www.merck.com/mrl/merckgeneindex.2.html)

The basic information flow for an overall gene identification protocol follows:

1. **Masking repetitive DNA:** A map of repeat locations shows where regulatory and protein-coding regions are unlikely to occur (Smit, 1996). It is therefore best to locate and remove interspersed and simple repeats from eukaryotic sequences as the first step in any gene identification analysis. Because such repeats rarely overlap promoters or the coding portions of exons, their locations can provide important negative information on the location of gene features. RepeatMasker (http://www.genome.washington.edu/analysistools/repeatmask.htm) provides annotation and masking of both interspersed and simple repeats.
2. **Database searches:** Sequence similarity to other genes or gene products provides strong positive evidence for exons. Thus,

searching for a known homologue is perhaps the most widely understood means of identifying new CDS. For protein-coding genes, translating the sequence in all six possible reading frames and using the result as a query against databases of amino acid sequences and functional motifs is usually the best first step for finding important matches. Once a homologue has been found, a program such as Procrustes http://www.hto.usc.edu/software/ procrustes) may be used to make an optimal alignment between the known gene product and the new gene. It is shown that:

a. Most ancient conserved regions, ACR (regions of protein sequences showing highly significant homologies across phyla), are already known and may be found in current databases.
b. Roughly 20–50% of newly found genes contain an ACR that is represented in the databases.
c. Rarely expressed genes are less likely to contain an ACR than moderately or highly expressed ones.

The EST databases probably contain fragments of a majority of all genes. Current estimates are that the public EST databases of human and mouse ESTs represent more than 80% of the genes expressed in these organisms. Thus they are an important resource for locating some part of most genes.

Finding good matches to ESTs is strong suggestion that the region of interest is expressed and can give some indication as to the expression profiles of the located genes. However, the problem of inferring function remains because an EST that represents a protein without known relatives will not be informative about gene function. An access to dbEST can be found at http://www.ncbi.nlm.nih.gov/dbEST/ index.html.

3. **Codon bias detection:** Statistical regularity suggesting apparent "codon bias" over a region is one of the indicators of protein-coding regions. Information on codon usage can be found at the Codon usage database, http://www.kazusa or. jp/codon/ or http://biochem.otago ac.nz: 800/Transterm/homepage.html. Most com–putational identification of protein-coding genes relies heavily on recognizing the somewhat diffuse regularities in protein-coding regions that are due to bias in codon usage. Some of the most informative coding measures are:
   - Dicodon counts, i.e. frequency counts for the occurrence of successive codon pairs
   - Certain measure of periodicity, i.e. the tendency of multiple occurrences of the same nucleotide to be found at distances of 3n bp
   - A measure of homogeneity versus complexity, i.e. counting long homopolymer runs
   - Open reading frame occurrence (Fickett and Tung, 1992). Many coding region detection programs are primarily the result of combining the numbers from one or more coding measures to form a single number called a discriminant. Typically, the discriminant is calculated in a sliding window (i.e. for successive subsequence of fixed length) and the result is plotted. To gain significant information from a coding measure discriminant, a DNA chain of more than 100 bases is required.

4. **Detecting functional sites in the DNA:** Because matches to template patterns may indicate the location of functional sites on the DNA, it would be more assuring if we are able to recognize the locations such as transcription factor binding sites and exon/intron junctions where the gene expression machinery

interacts with the nucleic acids. One way to summarize the essential information content of these locations termed signals is to give the consensus sequence, consisting of the most common base at each position of an alignment of specific binding sites. Consensus sequence are very useful as mnemonic devices but are typically not very reliable for discriminating true sites from pseudosites. One technique for discriminating between true sites and pseudosites with a basis in physical chemistry is the position weight matrix (PWM). A score is assigned to each possible nucleotide at each possible position of the signal. For any particular sequence, considered as a possible occurrence of the signal, the appropriate scores are summed to give a score to a potential site. Under some circumstances, this score may be approximately proportional to the energy of binding for a control ribonucleoprotein/protein.

Promoters: The promoter is an information-rich signal that regulates transcription, especially for eukaryotes. Computer recognition of promoters is important partly for the advance it may provide in gene identification. Available programs include those depending primarily on simple oligonucleotide frequency counts and those depending on libraries describing transcription factor binding specificities together with some description of promoter structure. Information on eukaryotic polymerase II (mRNA synthesis) promoters can be found from EPD at http://www.epd.isb-sib.ch.

Intron splice sites: PWMs have been complied for splice sites in a number of different taxonomic groups (Senapathy et al. 1990), and these may be the best resource available for analysis in many organisms. Unfortunately, PWM analysis of the splice junction provides rather low specificity perhaps because of the existence of multiple splicing mechanisms and regulated alternative splicing. Many of the integrated gene identification services provide separate splice site predictions. ExInt database (http://intron.bic.nus.edu.sg/exint/extint.html) provides information on the exton—intron structure of eukaryotic genes. Various information resources related to introns can be found at EID (http://mcb.harvard.edu/gilbert/EID/) and Intron (http://nutmeg.bio.indiana.edu/intron/index.html) servers.

Translation initiation site: In eukaryotes, if the transcription start site is known, and there is no intron interrupting the 5_ UTR, Kozak's rule (Kozak, 1996) probably will locate the correct initiation codon in most cases. Splicing is normally absent in prokaryotes, yet because of the existence of multicitronic operons, promoter location is not the key information. Rather, the key is reliable localization of the ribosome binding site. The TATA sequence about 30 bp from the transcription start site may be used as a possible resource.

Termination signals: The polyadenylation and translation termination signals also help to demarcate the extent of a gene.

Untranslated regions: The 5_ and 3_ UTRs are portions of the DNA/RNA sequences flanking the final coding sequence (CDS). They are highly specific both to the gene and to the species from which the sequences are derived. Thus UTRs can be used to locate CDS, e.g. eukaryotic polyadenylation signal (Wahle and Keller, 1996), mammalian pyrimidine-rich element (Baekelandt et al., 1997), mammalian AU-rich region (Senterre-Lesenfants et al., 1995), *E. coli* fbi (3_-CCGCGAAGTC-5_) element (French et al., 1997), and so on. Information on 5_ and 3_ UTRs of eukaryotic mRNAs can be found at UTRdb (http://bigrea.area.ba.cnr.it:8000/srs6/). Online bibliographies (indexed by years) on computational gene recognition are maintained at http://linkage.rockefeller.edu/wli/gene/right.html. Some computational gene discovery programs accessible through the WWW are listed in Table 6.1. Most programs are organism-specific and require a selection of organisms. Because no single prediction method is perfect, it is advisable to submit the query sequence to several different computational programs for the gene prediction.

Categories of Gene Prediction Programs

The current gene prediction methods can be classified into two major categories—abinitio-based and homology-based approaches.

Ab Initio-based Approach

It does so by relying on two major features associated with genes. The first is the existence of gene signals, which include start and stop codons, intron splice signals, transcription factor binding sites, ribosomal binding sites, and polyadenylation (poly-A) sites. In addition, the triplet codon structure limits the coding frame length to multiples of three, which can be used as a condition for gene prediction. The second feature used by ab initio algorithms is gene content, which is statistical description of coding regions. It has been observed that nucleotide composition and statistical patterns of the coding regions tend to vary significantly from those of the noncoding regions. The unique features can be detected by employing probabilistic models such as Markov models or hidden Markov models to help distinguish coding from noncoding regions.

Homology-based Approach

The homology-based method makes predictions based on significant matches of the query sequence with sequences of known genes. For instance, if a translated DNA sequence is found to be similar to a known protein or protein family from a database search, this can be strong evidence that the region codes for a protein. Alternatively, when possible exons of a genomic DNA region match a sequenced cDNA, this also provides experimental evidence for the existence of a coding region. Some algorithms make use of both gene-finding strategies. There are also a number of programs that actually combine predic-

Table 6.1: Web servers for gene identification

Program	*URL*
Aat	http://genome.cs.mtu.edu/aat.html
BCM GeneFinder	http://dot.imgen.bcm.tmc.edu:9331/gene-finder/gfb.html
CGG GeneFinder	http://genomic.sanger.ac.uk/gf/gfb.html
GeneID	http://www1.imim.es/software/geneid/index.html
GeneMark at EBI	http://www2.ebi.ac.uk/genemark/
Genie	http://www.fruitfly.org/seq—tools/genie.html
GenLang	http://www.cbil.upenn.edu/genlang/genlang—home.html
GenScan	http://genes.mit.edu/GENSCAN.html
Grail/Grail 2	http://compbio.ornl.gov/gallery.html
Mzef	http://www.cshl.org/genefinder/
ORFgene	http://www.ncbi.nlm.nih.gov/gorf/gorf.html
Procrustes	http://wwwhto.usc.edu/software/procrustes/wwwserv.html
WebGene	http://www.itba.mi.cnr.it/webgene/

tion results from multiple individual programs to derive a consensus prediction. This type of algorithms can therefore be considered as consensus-based.

Gene Prediction in Prokaryotes

Prokaryotes, which include bacteria and Archaea, have relatively small genomes with sizes ranging from 0.5 to 10Mbp(1Mbp=10^6 bp). The gene density in the genomes is high, with more than 90% of a genome sequence containing coding sequence. There are very few repetitive sequences. Each prokaryotic gene is composed of a single contiguous stretch of ORF coding for a single protein or RNA with no interruptions within a gene.

The following section describes a number of HMM/IMM-based gene finding programs for prokaryotic organisms.

GeneMark (http://opal.biology.gatech.edu/ GeneMark/) is a suite of gene prediction programs based on the fifth-order HMMs. The main program-GeneMark.hmm-is trained on a number of complete microbial genomes. If the sequence to be predicted is from a non-listed organism, the most closely related organism can be chosen as the basis for computation. Another option for predicting genes from a new organism is to use a self-trained program GeneMarkS as long as the user can provide at least 100 kbp of sequence on which to train the model. If the query sequence is shorter than 100 kbp, a GeneMark heuristic program can be used with some loss of accuracy. In addition to predicting prokaryotic genes, GeneMark also has a variant for eukaryotic gene prediction using HMM.

Glimmer (Gene locator and interpolated Markov modeler, www.tigr.org/softlab/glimmer/glimmer.html) is a UNIX program from TIGR that uses the IMM algorithm to predict potential coding regions. The computation consists of two steps, namely model building and gene prediction. The model building involves training by the input sequence, which optimizes the parameters of the model. In an actual gene prediction, the overlapping frames are "flagged" to alert the user for further inspection. Glimmer also has a variant, GlimmerM, for eukaryotic gene prediction.

FGENESB (www.softberry.com/berry.phtmltopic =gfindb) is a web-based program that is also based on fifth-order HMMs for detecting coding regions. The program is specifically trained for bacterial sequences. It uses the Vertibi algorithm to find an optimal match for the query sequence with the intrinsic model. A linear discriminant analysis (LDA) is used to further distinguish coding signals from non-coding signals.

These programs have been shown to be reasonably successful in finding genes in a genome. The common problem is imprecise prediction of translation initiation sites because of inefficient identification of ribosomal binding sites. This problem can be remedied by identifying the ribosomal binding site associated with a start codon. A number of algorithms have been developed solely for this purpose. RBSfinder is one such algorithm.

RBSfinder(ftp://ftp.tigr.org/pub/software/RBSfinder/) is a UNIX program that uses the prediction output from Glimmer and searches for the Shine-Delgarno sequences in the vicinity of predicted start sites. If a high-scoring site is found by the intrinsic probabilistic model, a start codon is confirmed; otherwise the program moves to other putative translation start sites and repeats the process.

Gene Prediction in Eukaryotes

Eukaryotic nuclear genomes are much larger than prokaryotic ones, with sizes ranging from 10 Mbp to 670 Gbp (1 Gbp = 10^9 bp). They tend to have a very low gene density. In humans, for instance, only 3% of the genome codes for genes, with about 1 gene per 100 kbp on average. The space between genes is often very large and rich in repetitive sequences and transposable elements. Most importantly, eukaryotic genomes are characterized by a mosaic organization in which a gene is split into pieces (called exons) by intervening noncoding sequences.

The main issue in prediction of eukaryotic genes is the identification of exons, introns, and splicing sites. From a computational point of view, it is a very complex and challenging problem. Because of the presence of split gene structures, alternative splicing, and very low gene densities, the difficulty of finding genes in such an environment is likened to finding a needle in a haystack. The needle to be found actually is broken into pieces and scattered in many different places. The job is to gather the pieces in the haystack and reproduce the needle in the correct order.

Gene Prediction Programs

To date, numerous computer programs have been developed for identifying eukaryotic genes. They fall into all three categories of algorithms: ab initio-based, homology-based, and consensus-based. Most of these programs are organism specific because training data sets for obtaining statistical parameters have to be derived from individual organisms. Some of the algorithms are able to predict the most probable exons as well as suboptimal exons providing information for possible alternative spliced transcription products.

Ab Initio Programs

The frequently used ab initio programs make use of neural networks, HMMs, and discriminant analysis, which are described next.

GRAIL (Gene Recognition and Assembly Internet Link; http://compbio.ornl.gov/public/tools/) is a web-based program that is based on a neural network algorithm. The program is trained on several statistical features such as splice junctions, start and stop codons, poly-A sites, promoters, and CpG islands. The program scans the query sequence with windows of variable lengths and scores for coding potentials and finally produces an output that is the result of exon candidates. The program is currently trained for human, mouse, Arabidopsis, Drosophila, and *Escherichia coli* sequences.

FGENES (FindGenes;www.softberry.com/) is a web-based program that uses LDA to determine whether a signal is an exon. In addition to FGENES, there are many variants of the program. Some programs, such as FGENESH, make use of HMMs. There are others, such as FGENESH C, that are similarity based. Some programs, such as FGENESH+, combine both ab initio and similarity-based approaches.

MZEF (Michael Zhang's Exon Finder; http://argon.cshl.org/genefinder/) is a web-based program that uses QDA for exon prediction. Despite the more complex mathematical functions, the expected increase in performance has not been obvious in actual gene prediction.

Prediction Using HMMs

GENSCAN(http://genes.mit.edu/GENSCAN.html) is a web-based program that makes predictions based on fifth-order HMMs. It combines hexamer frequencies with coding signals (initiation codons, TATA box, cap site, poly-A, etc.) in prediction. Putative exons are assigned a probability score (P) of being a true exon. Only predictions with $P > 0.5$ are deemed reliable. This program is trained for sequences from vertebrates, Arabidopsis, and maize. It has been used extensively in annotating the human genome.

HMMgene (www.cbs.dtu.dk/services HMM gene) is also an HMM-basedweb program.

The unique feature of the program is that it uses a criterion called the conditional maximum likelihood to discriminate coding from noncoding features. If a sequence already has a sub-region identified as coding region, which may be based on similarity with cDNAs or proteins in a database, these regions are locked as coding regions.

An HMM prediction is subsequently made with a bias towards the locked region and is extended from the locked region to predict the rest of the gene coding regions and even neighboring genes. The program is in a way a hybrid algorithm that uses both ab initio-based and homology-based criteria.

Homology-based Programs

Homology-based programs are based on the fact that exon structures and exon sequences of related species are highly conserved. When potential coding frames in a query sequence are translated and used to align with closest protein homologs found in databases, near perfectly matched regions can be used to reveal the exon boundaries in the query. This approach assumes that the database sequences are correct. It is a reasonable assumption in light of the fact that many homologous sequences to be compared with are derived from cDNA or expressed sequence tags (ESTs) of the same species. With the support of experimental evidence, this method becomes rather efficient in finding genes in an unknown genomic DNA. The drawback of this approach is its reliance on the presence of homologs in databases. If the homologs are not available in the database, the method cannot be used. Novel genes in a new species cannot be discovered without matches in the database.

GenomeScan(http://genes.mit.edu/genomescan. html) is a web-based server that combines GENSCAN prediction results with BLASTX similarity searches. The user provides genomic DNA and protein sequences from related species. The genomic DNA is translated in all six frames to cover all possible exons. The translated exons are then used to compare with the user-supplied protein sequences. Translated genomic regions having high similarity at the protein level receive higher scores. The same sequence is also predicted with a GENSCAN algorithm, which gives exons probability scores. Final exons are assigned based on combined score information from both analyses.

EST2Genome (http://bioweb.pasteur.fr/seqanal/interfaces/est2genome.html) is a web-based program purely based on the sequence alignment approach to define intron–exon boundaries. The program compares an EST (or cDNA) sequence with a genomic DNA sequence containing the corresponding gene. The alignment is done using a dynamic programming-based algorithm. One advantage of the approach is the ability to find very small exons and alternatively spliced exons that are very difficult to predict by any ab initio-type algorithms. Another advantage is that there is no need for model training, which provides much more flexibility for gene prediction. The limitation is that EST or cDNA sequences often contain errors or even introns if the transcripts are not completely spliced before reverse transcription.

SGP-1(SyntenicGenePrediction;http://195.37. 47.237/sgp-1/) is a similarity-based web program that aligns two genomic DNA sequences from closely related organisms. The program translates all potential exons in each sequence and does pairwise alignment for the translated protein sequences using a dynamic programming approach. The near-perfect matches at the protein level define coding regions. Similar to EST_2 Genome, there is no training needed. The limitation is the need for two homologous sequences having similar genes with similar exon structures; if this condition is not met, a gene escapes detection from one sequence when there is no counterpart in another sequence.

TwinScan (http://genes.cs.wustl.edu/) is also a similarity-based gene-finding server. It is similar to GenomeScan in that it uses Gen-

Scan to predict all possible exons from the genomic sequence. The putative exons are used for BLAST searching to find closest homologs. The putative exons and homologs from BLAST searching are aligned to identify the best match. Only the closest match from a genome database is used as a template for refining the previous exon selection and exon boundaries.

Consensus-based Programs

Because different prediction programs have different levels of sensitivity and specificity, it makes sense to combine results of multiple programs based on consensus. This idea has prompted development of consensus-based algorithms. These programs work by retaining common predictions agreed by most programs and removing inconsistent predictions. Such an integrated approach may improve the specificity by correcting the false positives and the problem of over prediction. However, since this procedure punishes novel predictions, it may lead to lowered sensitivity and missed predictions.

Two examples of consensus-based programs are given next.

GeneComber (www.bioinformatics.ubc.ca/genecomber/index.php) is a web server that combines HMMgene and GenScan prediction results. The consistency of both prediction methods is calculated. If the two predictions match, the exon score is reinforced. If not, exons are proposed based on separate threshold scores.

DIGIT (http://digit.gsc.riken.go.jp/cgi-bin/index.cgi) is another consensus-based web server. It uses prediction from three ab initio programs – FGENESH, GENSCAN and HMM gene. It first compiles all putative exons from the three gene-finders and assigns ORFs with associated scores. It then searches a set of exons with the highest additive score under the reading frame constraints. During this process, a Bayesian procedure and HMMs are used to infer scores and search the optimal exon set which gives the final designation of gene structure.

Integrated Gene Identification

The gene structure of an unknown nucleotide sequence can be identified by alignment against target protein sequences.

Procrustes (http://www-hto.usc.edu/software/procrustes/wwwserv.html) performs spliced alignment of the predicted protein translations of a query nucleotide sequence against the target protein sequences. On the query page, enter the name of the query DNA and paste the query nucleotide sequence, then enter the names of the target protein and paste the amino acid sequence of the target protein. Click the Run Procrustes button. The user may request to receive the results by e-mail or online. The online results are kept in http://www.hto.usc.edu/tmp/fileid.html for approximately 10 hours after completing the task. The report includes pairwise presentations of spliced alignment score, positions and length of exons, translation of predicted gene based on the target protein, and spliced alignment with the target protein.

GeneBuilder tool of WebGene (http://125.itba.mi.cnr.it/_webgene/genebuilder.-html) also derives the predicted gene from a query nucleotide sequence by comparing with an input protein sequence. Select (radio selections) prediction features (exon analysis, potential coding region identification, repeated element mapping, EST mapping, splice site prediction, potential coding regions, TATA box prediction, and poly-A site prediction) and click the Start analysis button. The GeneView tool of WebGene (http://125.itba.mit.cnr.it/_webgene/wwwgene.html) takes the query sequence and returns the predicted CDS in the GenBank format

The generalized hidden Markov model trained on human genes of Genie at http://www.fruitfly.org/seq—tools/genie.html accepts the input sequence in fasta format and returns the predicted exons by e-mail. The discriminant analysis at MZEF (http://argon.cshl.org/genefinder/human.htm) identifies protein coding regions in a human genome by returning the predicted internal coding exons (Zhang, 1997). The GeneFinder server (http://argon.cshl.org/genefinder/Pombe/pombe.htm) also predicts internal exons, introns, and ORF of a query sequence.

GeneID at http://www1.imim.es/software/geneid/index.html predicts splice sites, codons, and exons along the query sequence using PWM. Exons are built and scored as the sum of the scores of the sites, plus the log-likelihood ratio of a Markov model for coding DNA. The gene structure is then assembled from the assembled exons. On the Web server page of GeneID at http://www1.imim.es/geneid.html, paste the query sequence in fasta format (starting with Name of sequence), choose prediction options and output options (acceptor sites, donor sites, start codons, stop codons, first exons, internal exons, terminal exons, and single gene). Click the Submit button to display predicted gene structures in a text listing (positions, sequences, and scores) and a graphical presentation.

The discriminant analysis based on pattern recognition of GeneFinder is accessible at CGG of the Sanger Centre (http://genomics.sanger.ac.uk/gf/gfb.html) and Baylor College of Medicine (http://dot.imgen.bcm.tmc.edu:9331/gene-finder/gfb.html). The GeneFinder offers various programs for predicting functional sites (e.g. SPL, TSSG/TSSW, FEX, POLYAH and NSITE) and the gene structure such as FGENESH (hidden Markow model based gene structure prediction for multiple genes) and FGENE (finding gene by exon search and assembling). On the analysis request page of CGG at the Sanger Centre or BCM, paste the query sequence, enter the sequence name, choose an organism,select the program (FGENESH or FGENE), and click the Perform Search button.

The heuristic approach for the gene prediction of GeneMark can be accessed online at WebGeneMark (http://genemark.biology.gatech.edu/Gene-Mark/webgenemark.html) and at EBI (http://www2.ebi.ac.uk/genmark/). On the WebGeneMark query page, enter the sequence name, paste the query sequence, select running options and output options, and then click the Start GeneMark button. On the EBI GeneMark page, enter e-mail address, choose options (e.g. exons, ORFs, and regions in nucleotide—transcripts or protein—translations), paste or upload the query sequence, and click the Run GeneMark button. The query returns with lists of regions of interests (from start to stop codons), protein-coding exons (between acceptor and donor sites), and pointers to requested output files which can be viewed/saved.

GenScan combines information about gene feature statistics with a probabilistic model to predict gene structure. The web site is accessible at http://genes.mit.edu/GENSCAN.html. Paste the query sequence, select organism and print option and then click run GENSCAN button. The results (online or by e-mail) include clickable view of gene structure in pdf or ps images, a table listing predicted gene types (initial, internal and terminal exons, single-exon gene, promoter and poly-A signal), their length, beginning and end points, reading frames and scores of the predicted features.

Grail (Uberbacher, et al. 1993) combines a variety of signals and content information using a neural network to perform gene analysis. The gene recognition of Grail can be accessed at http://compbio.ornl.gov/Grail-1.3/Toolbar.html.

Click the New Grail Analysis menu on the tool bar to select an organism (human, mouse, Arabidopsis, or *E. coli*). Click Generate Request Form to open the query form. Select analysis options, paste the query sequence, and click the Submit Request button. After the completion of analysis, click Display Results Window to view/save the results with a listing of exons, prediction scores, and exon translations.

Web-based Tools

GENSCAN

GENSCAN, developed by Chris Burge and Sam Karlin, is designed to predict complete gene structures. As such, GENSCAN can identify introns, exons, promoter sites, and poly–A signals, as do a number of the other gene identification algorithms. Like FGENES, GENSCAN does not expect the input sequence to represent one and only one gene or one and only one exon: it can accurately make predictions for sequences representing either partial genes or multiple genes separated by intergenic DNA.

The ability to make these predictions accurately when a sequence is in a variety of contexts makes GENSCAN a particularly useful method for gene identification. GENSCAN relies on what the author terms a "probabilistic model" of genomic sequence composition and gene structure. By looking for gene structure descriptions that match or are consistent with the query sequence, the algorithm can assign a probability as to the chance that a given stretch of sequence represents an exon, promoter, and so forth. The "optimal exons" are the ones with the highest probability and represent the part of the query sequence having the best chance of actually being an exon. The method will also predict "suboptimal exons," stretches of sequence having an acceptable probability value but one not as good as the optimal one. The authors of the method encourage users to examine both sets of predictions so that alternatively spliced regions of genes or other nonstandard gene structures are not missed.

GRAIL

GRAIL, which stands for Gene Recognition and Analysis Internet Link, is the elder statesman of the gene prediction techniques because it is among the first of the techniques developed in this area and enjoys widespread usage. As more and more has become known about gene structure in general and better Internet tools have become more widespread, GRAIL has continuously evolved to keep in step with the current state of the field.

There are two basic GRAIL versions. GRAIL 1 makes use of a neural network method to recognize coding potential in fixed-length (100 base) windows considering the sequence itself, without looking for additional features such as splice junctions or start and stop codons. An improved version of GRAIL 1 (called GRAIL 1a) expands on this method by considering regions immediately adjacent to regions deemed to have coding potential, resulting in better performance in both finding true exons and eliminating false positives. Either GRAIL 1 or GRAIL 1a would be appropriate in the context of searching for single exons. A further refinement led to a second version, called GRAIL 2, in which variable-length windows are used and contextual information (e.g. splice junctions, start and stop codons, poly-A signals) is considered. Because GRAIL 2 makes its prediction by taking genomic context into account, it is appropriate for determining model gene structures.

GENE FINDER was developed to make sense of genomic DNA sequence data from *Aspergillus niger* and *Saccharomyces cerevisiae.* The program will only recognize genes beginning with ATG and ending with TAA, TAG or

TGA. If you wish to ensure detection of genes whose beginning or ending is not included in the submitted sequence simply add "ATG-CATGCATG" and "TAGCTAGCTAG" at the beginning and end of your sequence respectively. The program identifies and evaluates sequences conforming to the following consensus motifs:

1. Translation initiation site,
2. Translation stop site,
3. Left intron junction,
4. Right intron junction, and
5. Intron branch-point.

Potential introns are identified and assigned scores based on:

1. How well each of the sequences matches the consensus,
2. The spacing between the branch-point and the right junction, and
3. The length of the resulting intron.

The versions available here have been optimized for *Aspergillus niger* genomic DNA and *Saccharomyces cerevisiae.* Since most *A. niger* genes are interrupted by several (up to 8) introns, the *A. niger* program was designed to handle potential open reading frames (ORFs) containing up to 10 introns. The yeast program will identify ORFs having up to two introns. The following strategy is employed:

1. ORFs are constructed from each start site to the first in-frame stop codon.
2. These ORFs are then assigned a score based on their length.
3. Introns are sought which either excise the stop or alter the reading frame.
4. The next in-frame stop codon is identified.
5. The new ORF is given a score based on length and intron score.
6. And so on....

Since billions of hypothetical ORFs can be constructed from a few kilobases of *A. niger* genomic sequence, a few structures were applied. First, only the top 25 *A. niger* ORFs are stored in memory. For yeast, the top 100 ORFs are kept. Second, only potential introns which exceed a certain score are considered. (Reducing this stringency level results in a much slower program.)

Note also that the context of the start and stop site are not taken into consideration in this version of the program. This is mainly to streamline processing. Queries from the net are expected to be relatively small (kb not Mb). Information of DNA sequence up to 1 kb upstream of the start site are required to evaluate the likelihood of authenticity. However, in *A. niger* at least, most authentic start sites are easily identified.

Predicting Secondary Structure from Protein Sequences

Protein structure is intrinsically hierarchic in its internal organization. The highest level in this hierarchy is constituted by complete proteins or assemblies of such proteins, which become subdivided through domains via super-secondary structure to secondary structure at the lowest hierarchical level. At higher levels within in this hierarchy, especially from the domain level upwards, the connectivity of the polypeptide backbone between substructures becomes less important. A protein thus can retain a stable structure irrespective of the sequential arrangement of domains and presence of fragments linking them together. Such linker regions often constitute exposed surface loops that do not disrupt the folds of the domains they connect. At the level of protein secondary structure, however, the elements are not only crucially dependent on their amino acid compositions, but, unlike domain and higher-order structures, are also very much context dependent; i.e. they rely critically on the substructures in their environment. It is because of this context dependency, that predicting protein secondary structure is a very difficult task, which after three decades of research has not attained the accuracy on which further prediction of tertiary structure can be based.

It must be stressed, however, that some successful predictions of higher-order structure, based on a knowledge of the secondary structure, have been achieved.

What is Secondary Structure?

The two basic secondary structures are the α-helix and the β-strand. Both show distinct structural features and are easily recognizable in a protein structure. Other secondary structure types occurring in protein structures are more difficult to classify as they are less regular than α-helices or β-strands. Such structures are defined in the context of" most prediction methods as coil; i.e. leftover secondary structures that cannot be considered in the α-helical or β-stranded conformation.

In general, about 50% of the amino acids fold into α-helices or β-stands, so that roughly half the protein structures are irregularly shaped. The primary reason for the regularity observed for helices and strands is the inherent polar nature of the protein backbone, which contributes a polar nitrogen and oxygen atom for each amino acid. To satisfy energetical constraints, the parts of the main-chain buried in the internal protein core need to form hydrogen-bonds between those polar atoms. The α-helix and β-strand conformations are optimal as each main-chain nitrogen atom can associate with an oxygen partner (and vice versa) whenever they adopt one of these two secondary structure types. It must be stressed that, in order to satisfy their hydrogen-bonding constraints, β-strands need to interact with other β-strands, which they can do in a parallel and antiparallel fashion, thus forming a β-pleated sheet. β-strands thus depend on crucial long-range interactions between residues remote in sequence. They therefore are more context dependent than α-helices, which would be more able to fold 'on their own'. The fact that the vast majority of prediction methods have greatest difficulty in delineating β-strands correctly is believed to be due to their pronounced context dependency.

Where Could Knowledge about Secondary Structure Help?

Experimental evidence on early protein folding intermediates has shown that secondary structural elements form at early stages during the folding process. These results support the significance of the so-called 'framework' model of protein folding, where two or more secondary structural elements would associate early during folding to provide a structural frame to which subsequently other substructures could attach. Therefore, knowledge of protein secondary structural regions along the sequence is a prerequisite to model the folding process or kinetics associated with it. Also for tertiary model building, the ability to predict the secondary structure from the sequence alone is crucial, as it allows for docking experiments to be carried out on the predicted α-helices and β-strands.

On the architectural side of protein structure, it is possible to recognize the three dimensional topology by comparing the successfully predicted secondary structural elements of a query protein with a database of known topologies. Successful prediction here means parts of those helices and strands essential for the topology would have been predicted, without necessarily accurate prediction of the edges of those structures or the detection of non-essential secondary structures.

What Signals are There to be Recognized?

A number of observations on secondary structures as found in the large collection of protein structures deposited in the Protein Data Bank (PDB), could be summed up for each of the secondary structures α-helix, β-strand, and loop as follows.

α-helix:

a. As the number of residues per turn is 3.6 in the ideal case and helices are often positioned against a buried core, they have one phase contacting hydrophobic amino acids, while the other phase interacts with the solvent. Such amphipathic helices thus show a periodicity of three to four residues in hydrophobicity of the associated sequence stretch.
b. Proline residues do not occur in middle segments as they disrupt the α-helical turn. However, they are seen in the first two positions of α-helices.

β-strand

a. β-strands mostly fold into so-called β-pleated sheets which have two strands forming either edge. Therefore, the hydrophobic nature of edge strands is different from that of strands internal to a β-sheet. As side-chains of constituent residues along a β-strand alternate the direction in which they protrude, edge strands of β-sheets can show an alternating pattern of hydrophobic hydrophilic residues, while buried strands tend to contain merely hydrophobic residues.
b. As β-strand is the most extended conformation (i.e. consecutive Cα atoms are farthest apart), it takes relatively few residues to cross the protein core with a strand. Therefore, the number of residues in a β-strand is usually limited and can be anything from two or three amino acids, whereas helices shielding such strands from solvent comprise more residues.
c. The β-strands can be disrupted by single residues that induce a kink in the extended structure of the main-chain. Such so-called β-bulges are often comprised of relatively hydrophobic residues.

Coil

a. Multiple alignments of protein sequences often display gapped and/or highly variable regions, which would be expected to be associated with loop regions rather than the two basic secondary structures.
b. Loop regions contain a high proportion of small polar residues like Ala, Gly, Ser, and Thr. Glycine residues are seen in loop regions due also to their inherent flexibility.
c. Proline residues are often seen in loops as well. They are not observed in helices and strands as they kink the main-chain, although they can occur in the N-terminal two positions of α-helices as mentioned above.

In addition to the positional requirements in hydrophobicity, there are also general compositional differences between helix, strand and coil conformations and this is the signal used in many of the early prediction methods for single sequences. Methods that utilize multiple alignments can also exploit the fact that the amino acid exchange patterns are different for the three secondary structure states.

A few additional rules can help in clarifying the structure or function of a protein sequence, once the secondary structure is predicted:

a. Hydrophobic and particularly conserved hydrophobic residues are normally buried in the protein core.
b. More than 95% of all so-called β-α-β motifs; i.e. a β-strand followed in sequence by a α-helix and another β-strand, show a right-handed chirality.
c. Helices often cover up a core of β-strands. Therefore, if both α-helices and (β-strands are predicted, an attempt should be made

to distribute the helices evenly at either phase of a tentative β-sheet in topology modelling.

d. As mentioned, strictly conserved residues in different regions of a multiple alignment can be predicted with great confidence to be responsible for the catalytic functions, particularly if they are polar and predicted to be in loop structures hence unlikely to be buried. As active site residues are positioned together in a protein 3D structure, the coil structures they constitute should be brought together in a topology model.

Secondary Structure Prediction Servers

Jpred

Jpred(http://www.compbio.dundee.ac.uk/jpred) is a secondary structure prediction server powered by the Jnet algorithm. Jpred performs over 1000 predictions per week for users in more than 50 countries. Given either a single protein sequence or a multiple sequence alignment, Jpred derives alignment profiles from which predictions of secondary structure and solvent accessibility are made.

The predictions are presented as coloured HTML, plain text, PostScript, PDF and via the Jalview alignment editor to allow flexibility in viewing and applying the data. The new Jpred 3 server includes significant usability improvements that include clearer feedback of the progress or failure of submitted requests. Functional improvements include batch submission of sequences, summary results via e-mail and updates to the search databases. A new software pipeline will enable Jnet/Jpred to continue to be updated in sync with major updates to SCOP and UniProt and so ensures that Jpred 3 will maintain high-accuracy predictions.

NNPREDICT

NNPREDICT is a program that predicts the secondary structure type for each residue in an amino acid sequence. The basis of the prediction is a two-layer, feed-forward neural network. The network weights were determined by a separate program—a modification of the Parallel Distributed Programming suite of McClelland and Rumelhart.

NNPREDICT takes as input a sequence consisting of one-letter amino acid codes (A C D E F G H I K L M N P Q R S T V W Y) (NOTE: B and Z are not recognized as valid amino acid codes) or three-letter amino acid codes separated by spaces (ALA CYS ASP GLU PHE GLY HIS ILE LYS LEU MET ASN PRO GLN ARG SER THR VAL TRP TYR). The output is a secondary structure prediction for each position in the sequence. Multiple-chain proteins can be predicted either in pieces, or as a single sequence, with a '!' character between chains.

The predicted type will be either: 'H', a helix element; 'E', a beta strand element, or '-', a turn element. If your sequence contains any symbols that are not standard amino acids, '?'s will be used in the output to indicate that no prediction could be made in the window around the unrecognized amino acid. NNPREDICT uses the tertiary class of the protein (either none, all-alpha, all-beta, or alpha/beta) for prediction. The possible options are:

- Use no tertiary structure class for prediction (the default)
- Use all-alpha tertiary structure class for prediction
- Use all-beta tertiary structure class for prediction
- Use alpha/beta tertiary structure class for prediction

SOPMA

The secondary structure prediction called "SOPMA" which means self-optimised prediction from multiple alignment is available within

ANTHEPROT. SOPMA (self-optimized prediction method with alignment) is an improvement of SOPM method. These methods are based on the homologue method. The improvent takes place in the fact that SOPMA takes into account information from an alignment of sequences belonging to the same family.

Alternatively, and starting with release 4.7, you may submit a sequence to SOPMA from within ANTHEPROT through Internet on the PBIL NPS@ web server provided that the Netwok connection is activated. The SOPMA method is able to also predict the turn state but accuracies are given only for 3 states (helix, sheet, coil).

Principle of SOPM

This method has been checked against an updated release of the Kabsch and Sander database, 'DATABASE.DSSP', extended to 239 protein chains sharing less than 50% identity. The first step of the SOPM is to build-sub-databases of protein sequences and their known secondary structures drawn from 'DATABASE.DSSP' by: (i) making binary comparisons of all protein sequences and (ii) taking into account the prediction of structural classes of proteins. The second step is to submit each protein of the sub-database to a secondary structure prediction using a predictive algorithm based on sequence similarity. The third step is to iteratively determine the predictive parameters that optimize the prediction quality on the whole sub-database. The last step is to apply the final parameters to the query sequence. This new method correctly predicts 69% of amino acids for a three-state description of the secondary structure (alpha helix, beta sheet and coil) in the whole database (46,011 amino acids). The correlation coefficients are C alpha = 0.54, C beta = 0.50 and Cc = 0.48. Root mean square deviations of 10% in the secondary structure content are obtained.

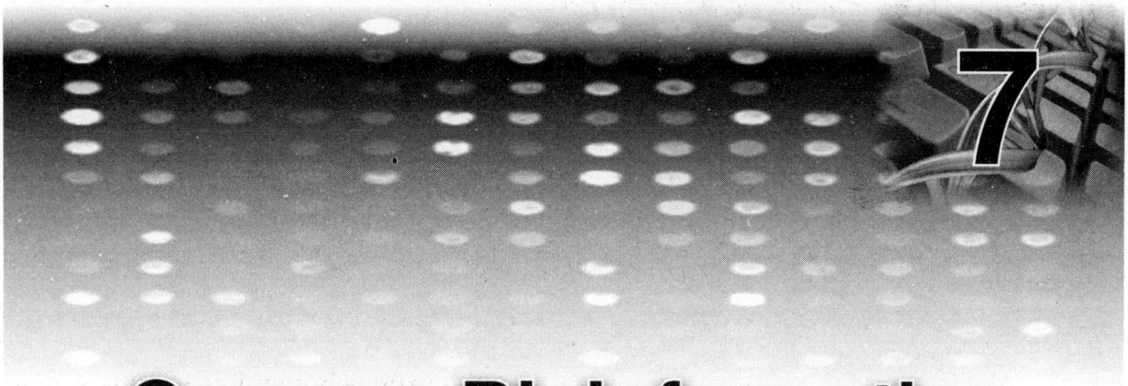

7 Genome Bioinformatics

Genomics is the study of genomes. Genomic studies are characterized by simultaneous analysis of a large number of genes using automated data gathering tools. The topics of genomics range from genome mapping, sequencing, and functional genomic analysis to comparative genomic analysis. The advent of genomics and the ensuing explosion of sequence information are the main driving force behind the rapid development of bioinformatics today.

Genomic study can be tentatively divided into structural genomics and functional genomics. Structural genomics refers to the initial phase of genome analysis, which includes construction of genetic and physical maps of a genome, identification of genes, annotation of gene features, and comparison of genome structures. However, it should to be mentioned that the term structural genomics has already been used by a structural biology group for an initiative to determine three-dimensional structures of all proteins in a cell. Strictly speaking, the initiative of structural determination of proteins falls within the realm of structural proteomics and should not be confused as a subdiscipline of genomics.

Genome Mapping

The first step in understanding a genome structure is through genome mapping, which is a process of identifying relative locations of genes, mutations or traits on a chromosome. A low-resolution approach to mapping genomes is to describe the order and relative distances of genetic markers on a chromosome. Genetic markers are identifiable portions of a chromosome whose inheritance patterns can be followed. For many eukaryotes, genetic markers represent morphologic phenotypes.

In addition to genetic linkage maps, there are also other types of genome maps such as physical maps and cytologic maps, which describe genomes at different levels of resolution. Their relations relative to the DNA sequence on a chromosome are illustrated in Fig. 7.1.

Genome Sequencing

There are two major strategies for whole genome sequencing — the shotgun approach and the hierarchical approach.

Shotgun Approach

In genetics, shotgun sequencing, also known as shotgun cloning, is a method used for sequencing long DNA strands. It is named by analogy with the rapidly-expanding, quasi-random firing pattern of a shotgun.

Since the chain termination method of DNA sequencing can only be used for fairly

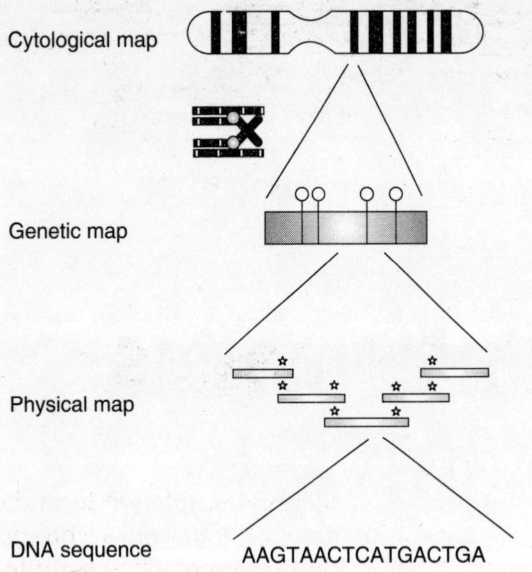

Fig. 7.1: Overview of various genome maps relative to the genomic DNA sequence. The maps represent different levels of resolution to describe a genome using genetic markers

short strands (100 to 1000 basepairs), longer sequences must be subdivided into smaller fragments, and subsequently re-assembled to give the overall sequence (Fig. 7.2).

Two principal methods are used for this: chromosome walking, which progresses through the entire strand, piece by piece, and shotgun sequencing, which is a faster but more complex process, and uses random fragments.

In shotgun sequencing, DNA is broken up randomly into numerous small segments, which are sequenced using the chain termination method to obtain reads. Multiple overlapping reads for the target DNA are obtained by performing several rounds of this fragmentation and sequencing. Computer programs then use the overlapping ends of different reads to assemble them into a continuous sequence .

Shotgun sequencing was one of the precursor technologies that was responsible for enabling full genome sequencing.

Example: Consider the following two rounds of shotgun reads:

Strand	Sequence
Original	AGCATGCTGCAGTCATGCTTAGGCTA
First shotgun	AGCATGCTGCAGTCATGCT-------
sequence	-------------------TAGGCTA
Second shotgun	AGCATG--------------------
sequence	------CTGCAGTCATGCTTAGGCTA
Reconstruction	AGCATGCTGCAGTCATGCTTAGGCTA

In this extremely simplified example, none of the reads cover the full length of the original sequence, however, the four reads can be assembled into the original sequence using the overlap of their ends to align and order them. In reality, this process uses enormous amounts of information that are rife with ambiguities and sequencing errors. Assembly of complex genomes is additionally com-

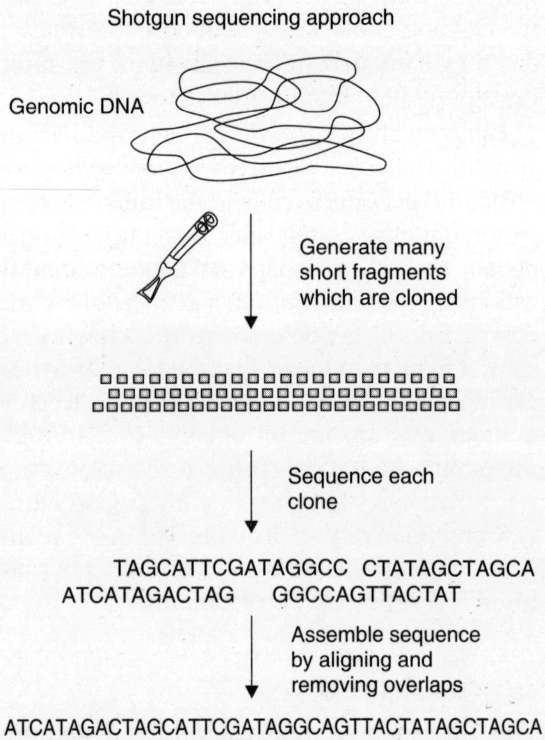

Fig. 7.2: Flow chart of shotgun approach of sequencing

plicated by the great abundance of repetitive sequence, meaning similar short reads could come from completely different parts of the sequence.

Hierarchical Approach

The hierarchical genome sequencing approach is similar to the shotgun approach, but on a smaller scale. The chromosomes are initially mapped using the physical mapping strategy. Longer fragments of genomic DNA (100 to 300 kB) are obtained and cloned into a high-capacity bacterial vector called bacterial artificial chromosome (BAC). Based on the results of physical mapping, the locations and orders of the BAC clones on a chromosome can be determined. By successively sequencing adjacent BAC clone fragments, the entire genome can be covered. The complete sequence of each individual BAC clone can be obtained using the shotgun approach. Overlapping BAC clones are subsequently assembled into an entire genome sequence (Fig. 7.3).

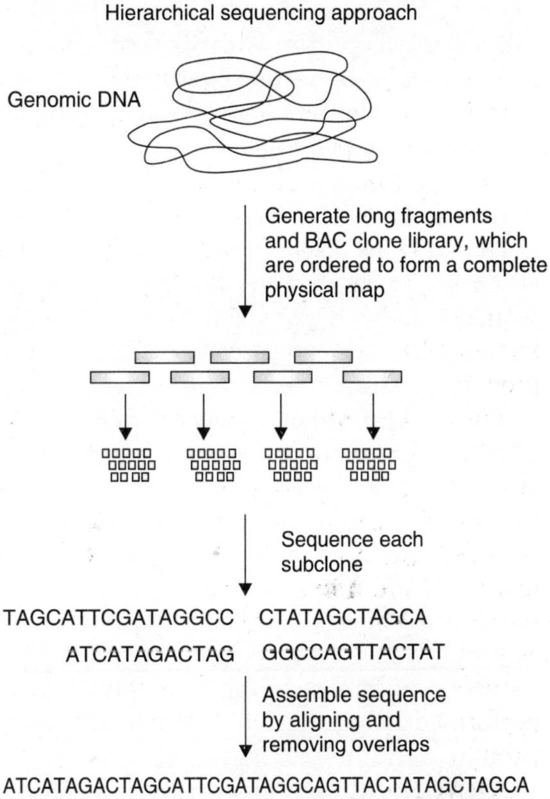

Fig. 7.3: Hierarchical approach of sequencing

Bioinformatics Tools and Automation in Genome Sequencing

Base Calling and Assembly Programs

The first step towards genome assembly is to derive base calls and assign associated quality scores. The next step is to assemble the sequence reads into contiguous sequences. This step includes identifying overlaps between sequence fragments, assigning the order of the fragments and deriving a consensus of an overall sequence. Assembling all shotgun fragments into a full genome is a computationally very challenging step. There are a variety of programs available for processing the raw sequence data. The following is a selection of base calling and assembly programs commonly used in genome sequencing projects.

Phred (www.phred.org/) is a UNIX program for base calling. It uses a Fourier analysis to resolve fluorescence traces and predict actual peak locations of bases. It also gives a probability score for each base call that may be attributable to error. The commonly accepted score threshold is twenty, which corresponds to a 1% chance of error. The higher the score, the better the quality of the sequence reads. If the score value falls below the threshold, human intervention is required.

Phrap (www.phrap.org/) is a UNIX program for sequence assembly. It takes Phred base-call files with quality scores as input and aligns individual fragments in a pairwise fashion using the Smith-Waterman algorithm. The base quality information is taken into account during the pairwise alignment. After all the pairwise sequence similarity is identified, the program performs assembly by progressively merging sequence pairs with decreasing similarity scores while removing overlapped regions. Consensus contigs are derived after joining all possible overlapped reads.

Vec Screen (www.ncbi.nlm.nih.gov/VecScreen/VecScreen.html) is a web-based program that helps detect contaminating bacterial vector sequences. It scans an input nucleotide sequence and compares it with a database of known vector sequences by using the BLAST program.

TIGR Assembler (www.tigr.org/) is a UNIX program from TIGR for assembly of large shotgun sequence fragments. It treats the sequence input as clean reads without consideration of the sequence quality. A main feature of the program is the application of the forward–reverse constraints to avoid misassembly caused by sequence repeats. The sequence alignment in the assembly stage is performed using the Smith-Waterman algorithm.

ARACHNE(www-genome.wi.mit.edu/wga/) is a free UNIX program for the assembly of whole-genome shotgun reads. Its unique features include using a heuristic approach similar to FASTA to align overlapping fragments, evaluating alignments using statistical scores, correcting sequencing errors based on multiple sequence alignment, and using forward–reverse constraints. It accepts base calls with associated quality scores assigned by Phred as input and produces scaffolds or a fully assembled genome.

EULER (http://nbcr.sdsc.edu/euler/) is an assembly algorithm that uses a Eulerian superpath approach, which is a polynomial algorithm for solving puzzles such as the famous "traveling salesman problem": finding the shortest path of visiting a given number of cities exactly once and returning to the starting point. In this approach, a sequence fragment is broken down to tuples of twenty nucleotides. The tuples are distributed in a diagram with numerous nodes that are all interconnected. The tuples are converted to binary vectors in the nodes. By using a Viterbi algorithm, the shortest path among the vectors can be found, which is the best way to connect the tuples into a full sequence. Because this approach does not directly rely on detecting overlaps, it may be advantageous in assembling sequences with repeat motifs.

Genome Annotation

With the rapid accumulation of genomic sequence information, there is a pressing need to use computational approaches to accurately predict gene structure. Computational gene prediction is a prerequisite for detailed functional annotation of genes and genomes. The process includes detection of the location of open reading frames (ORFs). An ORF is a length of DNA sequence that contains a contiguous set of codons, each of which specifies an amino acid. The ultimate goal is to describe all the genes computationally with near 100% accuracy.

Comparative Genomics

Comparison of whole genomes from different organisms is comparative genomics, which includes comparison of gene number, gene location, and gene content from these genomes. The comparison helps to reveal the extent of conservation among genomes, which will provide insights into the mechanism of genome evolution and gene transfer among genomes. It helps to understand the pattern of acquisition of foreign genes through lateral gene transfer. It also helps to reveal the core set of Gene's common among different genomes, which should correspond to the genes that are crucial for survival. This knowledge can be potentially useful in future metabolic pathway engineering.

The main themes of comparative genomics include whole genome alignment, comparing gene order between genomes, constructing minimal genomes, and lateral gene transfer among genomes.

Whole Genome Alignment

With an ever-increasing number of genome sequences available, it becomes imperative to understand sequence conservation between genomes, which often helps to reveal the presence of conserved functional elements. This can be accomplished through direct genome comparison or genome alignment.

MUMmer (maximal unique match, www.tigr.org/tigrscripts/CMR2/webmum/mumplot) is a free UNIX program from TIGR for alignment of two entire genome sequences and comparison of the locations of orthologs. The program is essentially a modified BLAST, which, in the seeding step, finds the longest approximate matches that include mismatches instead of finding exact k-mer matches as in regular BLAST. The result of the alignment of whole genomes is shown as a dot plot with lines of connected dots to indicate collinearity of genes. It is optimized for pairwise comparison of closely related microbial genomes.

BLASTZ (http://bio.cse.psu.edu/) is a UNIX program modified from BLAST to do pairwise alignment of very large genomic DNA sequences. The modified BLAST program first masks repetitive sequences and searches for closely matched "words," which are defined as twelve identical matches within a stretch of nineteen nucleotides. The words serve as seeds for extension of alignment in both directions until the scores drop below a certain threshold. Nearby aligned regions are joined by using a weighted scheme that employs a unique gap penalty scheme that tolerates minor variations such as transitions in the seeding step of the alignment construction to increase its sensitivity.

LAGAN (limited area global alignment of nucleotides; http://lagan.stanford.edu/) is a web-based program designed for pairwise alignment of large genomes. It first finds anchors between two genomic sequences using an algorithm that identifies short, exactly matching words. Regions that have high density of words are selected as anchors. The alignments around the anchors are built using the Needleman–Wunsch global alignment algorithm. Nearby aligned regions are further connected using the same algorithm. The unique feature of this program is that it is able to take into account degeneracy of the genetic codes and is therefore able to handle more distantly related genomes. Multi-LAGAN, an extension of LAGAN, available from the same website, performs multiple alignment of genomes using a progressive approach similar to that used in Clustal.

PipMaker (http://bio.cse.psu.edu/cgi-bin/pip maker) is a web server using the BLASTZ heuristic method to find similar regions in two DNA sequences. It produces a textual output of the alignment result and also a graphical output that presents the alignment as a percent identity plot as well as a dot plot. For comparing multiple genomes, Multi PipMaker is available from the same site.

MAVID (http://baboon.math.berkeley.edu/mavid/) is a web-based program for aligning multiple large DNA sequences. MAVID is based on a progressive alignment algorithm similar to Clustal. It produces an NJ tree as a guide tree. The sequences are aligned recursively using a heuristic pairwise alignment program called AVID. AVID works by first selecting anchors using the Smith–Waterman algorithm and then building alignments for the sequences between nearby anchors. Connected alignments are treated as new anchors for building longer alignments. The process is repeated iteratively until the entire sequence pair including weakly conserved regions are aligned.

GenomeVista (http://pipeline.lbl.gov/cgi-bin/GenomeVista) is a database searching program that searches against the human, mouse, rat, or Drosophila genomes using a large piece of DNA as query. It uses a program called BLAT to find anchors and extends the alignment from the anchors using AVID. (BLAT is a fast local alignment algorithm that aligns short sequences of forty bases with more than 95% similarity.) It produces a graphical output that shows the sequence percent identity.

Finding a Minimal Genome

One of the goals of genome comparison is to understand what constitutes a minimal genome, which is a minimal set of genes required for maintaining a free-living cellular organism. Finding minimal genomes helps provide an understanding of genes constituting key metabolic pathways, which are critical for a cell's survival. This analysis involves identification of orthologous genes shared between a number of divergent genomes.

Coregenes (http://pasteur.atcc.org:8050/CoreGenes1.0//) is a web-based program that determines a core set of genes based on comparison of four small genomes. The user supplies NCBI accession numbers for the genomes of interest. The program performs an iterative BLAST comparison to find orthologous genes by using one genome as a reference and another as a query. This pairwise comparison is performed for all four genomes. As a result, the common genes are compiled as a core set of genes from the genomes.

Lateral Gene Transfer

Lateral gene transfer (or horizontal gene transfer) is defined as the exchange of genetic materials between species in a way that is incongruent with commonly accepted vertical evolutionary pathway. Lateral gene transfer mainly occurs among prokaryotic organisms when foreign genes are acquired through mechanisms such as transformation (direct uptake of foreign DNA from environment), conjugation (gene uptake through mating behavior), and transduction (gene uptake mediated by infecting viruses). The transmission of genes between organisms can occur relatively recently or as a more ancient event.

If lateral transfer events occurred relatively recently, one would expect to discover traces of the transfer by detecting regions of genomic sequence with unusual properties compared to surrounding regions. The unusual characteristics to be examined include nucleotide composition, codon usage, and amino acid composition. This can be considered a "within-genome" approach. Another way to discern lateral gene transfer is through phylogenetic analysis referred to as an "among-genome" approach, which can be used to discover both recent and ancient lateral gene transfer events.

ACT (Artemis comparison tool; www.sanger.ac.uk/Software/ACT) is a pairwise genomic DNA sequence comparison program (written in Java and run on UNIX, Macintosh, and Windows) for detecting gene insertions and deletions among related genomes. The pairwise sequence alignment is conducted using BLAST. The display feature includes showing collinear as well as noncollinear (rearrangement) regions between two genomes. It also calculates GC biases to indicate nucleotide patterns. However, it is up to the genome annotators to determine whether the observations constitute evidence for lateral gene transfer, as this requires combining evidence from multiple approaches.

Gene Order Comparison

Another aspect of comparative genomics is the comparison of gene order. When the order of a number of linked genes is conserved between genomes, it is called synteny. Generally speaking, gene order is much less conserved compared with gene sequences. Gene order conservation is in fact rarely observed among divergent species. Therefore, comparison of syntenic relationships is normally carried out between relatively close lineages. However, if syntenic relationships for certain genes are indeed observed among divergent prokaryotes, they often provide important clues to functional relationships of the genes of interest. For example, genes involved in the same metabolic pathway tend to be clustered among phylogenetically diverse organisms. The preservation of the gene order is a result of the selective pressure to allow the genes to be coregulated and function as an operon. Furthermore, the synteny of genes from divergent groups often associates with physical interactions of the encoded gene products.

GeneOrder (http://pumpkins.ib3.gmu.edu:8080/geneorder/) is a web-based program that allows direct comparison of a pair of genomic sequences of less than 2 Mb. It displays a dot plot with diagonal lines denoting collinearity of genes and lines off the diagonal indicating inversions or rearrangements in the genomes.

Bioinformatics Tools in Microarray Data Analysis

The most commonly used global gene expression profiling method in current genomics research is the DNA microarray-based approach. A microarray (or gene chip) is a slide attached with a high-density array of immobilized DNA oligomers (sometimes cDNAs) representing the entire genome of the species under study. Below mentioned are list of few bioinformatics tools which will be useful in microarray data analysis.

AMP: [automated microarray pipeline) (http://www.tm4.org/amp.html)] is a web-based application that allows the normalization of microarray data. AMP users can upload data in the form of Affymetrix CEL files, and define an analysis pipeline by selecting several intuitive options on the following pages. AMP will performs data normalization (e.g. RMA), basic statistical analysis (e.g. t-test, ANOVA), and analysis of annotation using gene classification (e.g. Gene Ontology term assignment). The analyses are performed without user intervention and the results are presented in a web-based summary that allows data to be downloaded in a variety of formats compatible with further directed analysis.

MADAM: [microarray data manager (http: //www.tm4.org/madam.html)]: The Microarray Data Manager, implemented in Java, facilitates the entry of data into a relational database. MADAM guides users through the microarray process from RNA procurement to data analysis, offering intelligent forms to simplify the tracking of experimental parameters and results that are essential for the interpretation of expression results in downstream analyses. Canned reports provide information on RNA samples, studies, slide maps and other pertinent data and a general SQL query window allows free form access to the underlying database. MADAM also serves as a platform for launching other data entry and management tools. Through the use of these integrated modules, users can view and score PCR plates, design experiments and studies, and track laboratory materials. Although not yet supported, MADAM is being adapted to read and write MAGE-ML, the XML data exchange format being developed by an international consortium of leading public databases and microarray research centers. A MAGE-ML version of MADAM facilitates submission of microarray data to public repositories such as Array Express and GEO.

Managing array data effectively requires development and maintenance of a database designed to reflect the experimental process. Central to TM4 is a MySQL database that stores all data related to the microarray process, such as information on experiments, studies, protocols, data normalization, and gene expression.

This database supports MIAME (Minimal Information About a Microarray Experiment; (Brazma et al., 2001)) and captures the information deemed essential by this standard. While this may serve as a complete microarray database system for some users, others can use it as a starting point for customized database development. A schema for the database is provided and TM4 can be adapted to work with existing relational databases as the interaction is via JDBC (Sun Microsystem's Java DataBase Connectivity Application Program Interface).

MIDAS: [microarray data analysis system (http://www.tm4.org/midas.html)]: Before the intensity values measured in TIGR Spotfinder can be compared, normalization is necessary. This critical step can help compensate for variability between slides and fluorescent dyes, as well as other systematic sources of error, by appropriately adjusting the measured array intensities. Data filtering can reduce the dataset by removing poor or questionable data, in addition to data deemed uninteresting or irrelevant to the analysis. TIGR's microarray data analysis system, a Java application, provides users an intuitive interface to design analysis protocols combining one or more normalization and filtering steps. In this way, data from many individual hybridizations can be treated in a uniform and reproducible manner. MIDAS reads ". tav" files generated by TIGR Spotfinder or retrieved from the database via MADAM. Normalization modules include locally weighted linear regression [lowess; (Cleveland and Devlin, 1988; Yang et al., 2002)] and total intensity normalization. These can be linked with filters, including low-intensity cutoff, intensity-dependent Z-score cutoffs, and replicate consistency trimming, creating a highly customizable method for preparing expression data for subsequent comparison and analysis. Data analysis methods are constructed using an intuitive graphical scripting language and can be saved for application to other datasets. MIDAS provides scatter plots that illustrate the effects of each algorithm on the data. When the normalization and filtering steps are complete, MIDAS outputs the data in tav format.

TM4 Utilities (http://www.tm4.org/utilities.html) TM4 software group also developed a variety of other useful tools. They are all distributed under the Artistic License and available for free download.

ExpressConverter

ExpressConverter is a file transformation tool that reads microarray data files in a variety of file formats and generates TIGR MultiExperiment Viewer file (.mev) as output so that the microarray data can be uploaded to the database with MADAM and analyzed with MIDAS and MEV.

SlideMap

SlideMap.pm is a Perl module used to create a row/column to plate/well map of a microarray slide. It does not as yet incorporate annotation data into the map, but simply creates the map object based on input parameters. The map is an ordered list of spots containing array_row, array_col, plate_alias, well. SlideMap currently supports 5 types of arrayers: IAS, MD, MD3, Lucidia, and Stanford. However, 'generic' arrayers can be used with this Perl module. SlideMap provides methods for conversion of spots to wells and wells to spots.

MeV: [MultiExperiment viewer (http://www.tm4.org/mev.html)]: Normalized and filtered expression files can be analyzed using TIGR Multiexperiment Viewer (MeV). MeV is a versatile microarray data analysis tool, incorporating sophisticated algorithms for clustering, visualization, classification, statistical analysis and biological theme discovery. MeV can handle several input file formats.

These include the ".mev" and ".tav" files generated by TIGR Spotfinder and TIGR MIDAS, and also Affymetrix® (".txt") and Genepix® (".gpr") files. MeV generates informative and interrelated displays of expression and annotation data from single or multiple experiments. At this final stage of the TM4 pipeline, flexibility and the variety of analysis techniques are critical, as every algorithm has strengths that can be exploited when used on certain datasets and experimental designs. The concept of modularization lends itself particularly well to this system, as novel algorithms and existing codebases from the microarray community can be integrated with the Java-based MeV using a well-defined module API.

Many of the algorithms currently implemented are listed below. Boot-strapping, jackknifing and k-means support resample the dataset to generate consensus clusters. Modules to allow metabolic pathways and genomic/chromosomal maps to be viewed with expression data overlaid are in development and testing. A wizard to handle links to public database websites is also being developed. Clusters identified through any analysis method can be labeled and tracked through other analyses, providing the user with the ability to compare the results of several clustering algorithms to determine consensus and focus on genes with specified expression patterns and biological profiles.

OligoWiz (www.cbs.dtu.dk/services/OligoWiz/) is a Java program that runs locally but allows the user to connect to the server to perform analysis via a graphic user interface. It designs oligonucleotides by incorporating multiple criteria including homology, Tm, low complexity, and relative position within a transcript.

OligoArray (http://berry.engin.umich edu /oligoarray2/) is also a Java client-server program that computes oligonucleotides for microarray construction. It uses the normal criteria with an emphasis on gene specificity and secondary structure for oligonucleotides. The secondary structures and related thermodynamic parameters are calculated using Mfold.

ArrayDB (http://genome.nhgri.nih.gov/arraydb/) is a web interface program that allows the user to upload data for graphical viewing. The user can present histograms, select actual microarray slide images, and display detailed information of each spot which is linked to functional annotation of the corresponding gene in the UniGene, Entrez, dbEST, and KEGG databases. This can help to provide a synopsis of gene function when interpreting the microarray data.

ScanAlyze (http://rana.lbl.gov/EisenSoftware.htm) is a windows program for microarray fluorescent image analysis. It features semi-automatic spot definition and multichannel pixel and spot analyses.

TIGR Spotfinder (http://www.tigr.org/softlab/) is another windows program for microarray image processing using the TIFF image format. It uses an adaptive threshold algorithm, which resolves the boundaries of spots according to their shapes. The algorithm determines the intensity of irregular spots more accurately than most other similar programs. It also interfaces with a gene expression database.

7

Arrayplot (www.biologie.ens.fr/fr/genetiqu/puces/publications/arrayplot/index.html) is a windows program that allows visualization, filtering, and normalization of raw microarray data. It has an interface to view significantly up-regulated or down-regulated genes. It calculates normalization factors based on the overall median signal intensity.

SNOMAD (http://pevsnerlab.kennedyk rieger. org/snomadinput.html) is a web server for microarray data normalization. It provides scatter plots based on raw signal intensities and performs log-transformation and linear regression as well as Lowess regression analysis of the data.

MA-ANOVA (www.jax.org/staff/churchill/labsite/software/anova/) is a statistical program for windows and UNIX that uses ANOVA to analyze microarray data. It calculates log ratios, displays ratio-intensity plots, and performs permutation tests and boot-strapping of confidence values.

Cyber-T (http://visitor.ics.uci.edu/genex/cybert/) is a web server that performs t-tests on observed changes of replicate gene expression measurements to identify significantly differentially expressed genes. It also contains a computational method for estimating false-positive and false-negative levels in experimental data based on modelling of P-value distributions.

Expressed Sequence Tag

An expressed sequence tag or EST is a short sub-sequence of a transcribed cDNA sequence. They may be used to identify gene transcripts, and are instrumental in gene discovery and gene sequence determination.

An EST is produced by one-shot sequencing of a cloned mRNA (i.e. sequencing several hundred base pairs from an end of a cDNA clone taken from a cDNA library). The resulting sequence is a relatively low quality fragment whose length is limited by current technology to approximately 500 to 800 nucleotides. Because these clones consist of DNA that is complementary to mRNA, the ESTs represent portions of expressed genes. They may be present in the database as either cDNA/mRNA sequence or as the reverse complement of the mRNA, the template strand.

ESTs can be mapped to specific chromosome locations using physical mapping techniques, such as radiation hybrid mapping or FISH. Alternatively, if the genome of the organism that originated the EST has been sequenced one can align the EST sequence to that genome.

Sources of Data and Annotations

dbEST

The rapid accumulation of EST sequences has prompted the establishment of public and private databases to archive the data. For example, GenBank has a special EST database established in 1992, dbEST (www.ncbi.nlm.nih.gov/dbEST/) that contains EST collections for a large number of organisms (>250). The database is regularly updated to reflect the progress of various EST sequencing projects. Each newly submitted EST sequence is subject to a database search. If a strong similarity to a known gene is found, it is annotated accordingly.

EST Index Construction

One of the goals of the EST databases is to organize and consolidate the largely redundant EST data to improve the quality of the sequence information so the data can be used to extract full-length cDNAs. The process includes a preprocessing step that removes vector contaminants and masks repeats. This is followed by a clustering step that associates EST sequences with unique genes. The next step is to derive consensus sequences by fusing redundant, overlapping ESTs and to correct errors, especially frameshift errors. This step results in longer EST contigs. The procedure is somewhat similar to the genome assembly of shotgun sequence reads. Finally, the coding regions are defined through the use of HMM-based gene finding algorithms.

To go another step further, compiled ESTs can be used to align with the genomic sequence if available to identify the genome locus of the expressed gene as well as intron–exon boundaries of the gene.

SIM4 (http://pbil.univ-lyon1.fr/sim4.php). The clustering process that reduces the EST redundancy and produces a collection of nonredundant and annotated EST sequences is known as gene index construction.

UniGene (www.ncbi.nlm.nih.gov/UniGene/) is an NCBI EST cluster database.

Each cluster is a set of overlapping EST sequences that are computationally processed to represent a single expressed gene. The database is constructed based on combined information from dbEST, GenBankmRNAdatabase, and" electronically spliced" genomic DNA. Only ESTs with 3´ poly-A ends are clustered to minimize the the problem of chimerism. The resulting 3´ EST sequences provide more unique representation of the transcripts. The next step is to remove contaminant sequences that include bacterial vectors and linker sequences. The cleaned ESTs are used to search against a database of known unique genes (EGADdatabase) with the BLAST program. The compiling step identifies sequence overlaps and derives sequence consensus using the CAP3 program. During this step, errors in individual ESTs are corrected; the sequences are then partitioned into clusters and assembled into contigs. The final result is a set of nonredundant, gene-oriented clusters known as UniGene clusters. Each UniGene cluster represents a unique gene and is further annotated for putative function and its gene locus information, as well as information related to the tissue type where the gene has been expressed. The entire clustering procedure is outlined in Fig. 7.4.

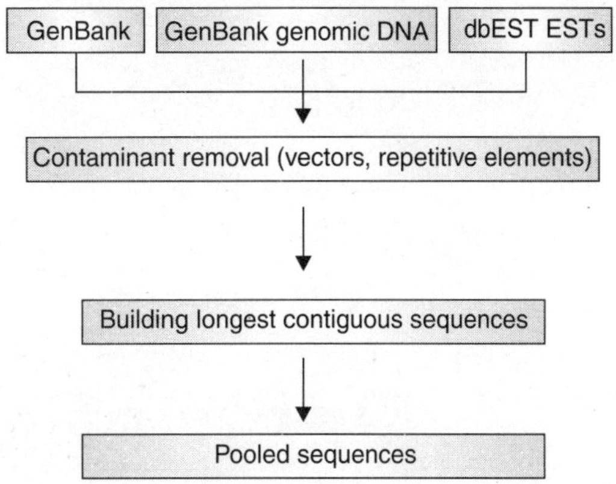

Fig. 7.4: Outline of steps to process EST sequences for construction of the UniGene database

ESTs and Sequence Polymorphisms

Single nucleotide polymorphisms (SNPs) can help to associate sequence variations with heritable phenotypes, facilitate studies in population and evolutionary biology, and aid in positional cloning and physical mapping. On average, SNPs occur every 500–1,000 nucleotides in human DNA. Gene-associated SNPs are found in untranslated regions as well as coding sequences (cSNPs). Because ESTs are sequenced redundantly from libraries prepared from different individuals, they seem an ideal source of polymorphic data. Indeed, a number of recent studies demonstrate that analysis of aligned EST sequences can lead to SNP discovery. All rely on alignment of EST sequences, identification of sequence differences, and a method to distinguish real polymorphisms from base calling (sequencing) errors and other artifacts. The public database for SNPs, dbSNP, is maintained at the NCBI. dbSNP accepts submissions not only of single nucleotide polymorphisms but of other polymorphisms, such as short deletions and inserts, microsatellites, and polymorphic insertion elements like retrotransposons. dbSNP is integrated with other NCBI resources such as GenBank, PubMed, genome sequences, and LocusLink.

8 Restriction Mapping

Restriction mapping is the process of obtaining structural information on a piece of DNA by the use of restriction enzymes.

Restriction Enzymes

Restriction enzymes are enzymes that cut DNA at specific recognition sequences called "sites." They probably evolved as a bacterial defense against DNA bacteriophage. DNA invading a bacterial cell defended by these enzymes will be digested into small, non-functional pieces. The name "restriction enzyme" comes from the enzyme's function of restricting access to the cell. A bacterium protects its own DNA from these restriction enzymes by having another enzyme present that modifies these sites by adding a methyl group. For example, *E. coli* makes the restriction enzyme Eco RI and the methylating enzyme Eco RI methylase. The methylase modifies Eco RI sites in the bacteria's own genome to prevent it from being digested.

Restriction enzymes are endonucleases that recognize specific 4 to 8 base regions of DNA. For example, one restriction enzyme, Eco RI, recognizes the following six bases sequence:

5' . . . G-A-A-T-T-C . . . 3'
3' . . . C-T-T-A-A-G . . . 5'

A piece of DNA incubated with Eco RI in the proper buffer conditions will be cut wherever this sequence appears. As you can see, this site is palindromic; that is, reading the upper strand from 5' to 3' is the same as reading the lower strand from 5' to 3'. As a result, each strand of the DNA can self-anneal and the DNA forms a small cruciform structure (Fig. 8.1).

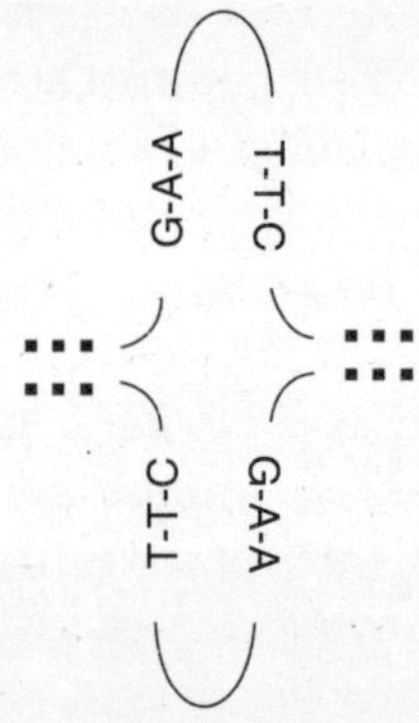

Fig. 8.1: Self-annealing in DNA

All restriction enzyme sites are palindromic. This structure may help the enzyme to recognize the sequence that it is designed to cut.

There are hundreds of restriction enzymes that have been isolated and each one recognizes its own specific nucleotide sequence. Sites for each restriction enzyme are distributed randomly throughout a particular DNA stretch.

Digestion of DNA by restriction enzymes is very reproducible; every time a specific piece of DNA is cut by a specific enzyme, the same pattern of digestion will occur. Restriction enzymes are commercially available and their use has made manipulating DNA very easy.

Restriction Mapping

Restriction mapping involves digesting DNA with a series of restriction enzymes and then separating the resultant DNA fragments by agarose gel electrophoresis. The distance between restriction enzyme sites can be determined by the patterns of fragments that are produced by the restriction enzyme digestion. In this way, information about the structure of an unknown piece of DNA can be obtained. An example of how this works is shown below. You have isolated a clone in pBluescript. You know how big the pBluescript portion of the plasmid is (3.0 kilobases) and what restriction enzymes are present in the plasmid (because you have its restriction map from the company that sold you the plasmid). You also know that the insert is 2.0 kb long and that it is inserted the Eco RI site. Your task is to find out more information about the insert.

At this point, you would digest plasmid with an enzyme that you know is in the pBluescript plasmid. For example, you know that there is only one Bam HI site in pBluescript, and it is in the multiple cloning site next to the Eco RI site (Fig. 8.2). If you digest this plasmid with Bam HI, there are two possibilities:

1. There are no Bam HI sites in the insert. If this is the case, when you run this digestion on a gel you will see only one DNA fragment, and it will be 5.0 kb long (3.0 kb of pBluescript DNA and 2.0 kb of insert DNA).
2. There is a Bam HI site in the insert. If this is the case, then the enzyme will cut the circular plasmid in two places, in the pBluescript part of the plasmid and in the insert. In this case, you will end up with two fragments of DNA. One will be pBluescript with some of the insert still attached and the other will be just insert. The sizes of the two fragments (determined by electrophoresis) will tell you where the site is. These two possibilities are shown in Fig. 8.3:

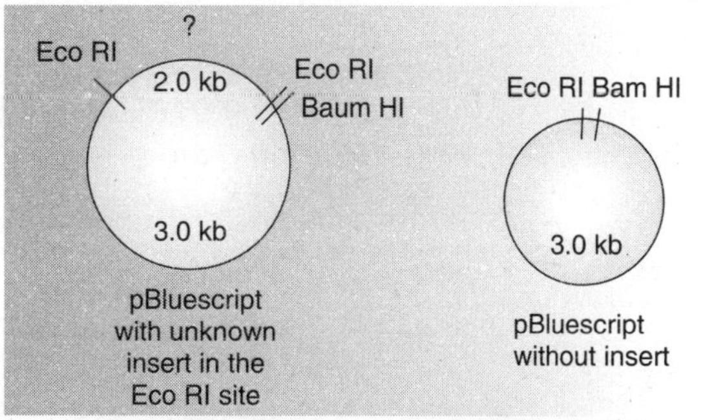

Fig. 8.2: Restriction map of pBluescript

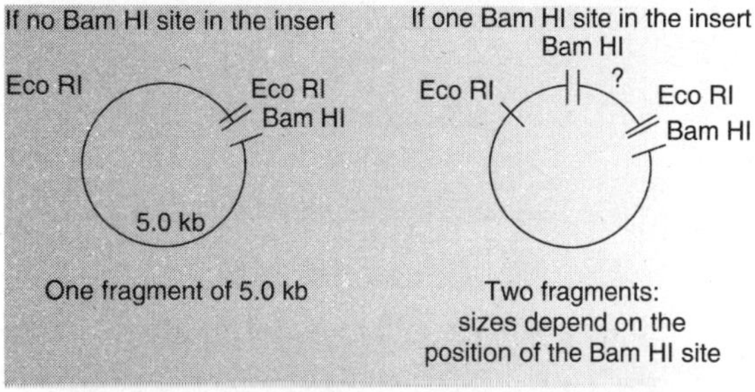

Fig. 8.3: Possibilities of restriction digestion

In the second case, where there is a site in the insert, the gel might look like in Fig. 8.4.

In this case, we learn two pieces of information:

1. That there is a Bam HI site in the insert, and
2. Where the site is in relation to the one end of the insert. When the Bam HI digestion is separated on an agarose gel, the sizes of the two fragments can be determined. In the above gel, the fragments are 3.6 kb and 1.4 kb. Therefore, we know that the Bam HI site is 1.4 kb away from the right hand side of the insert (Fig. 8.5). In this way, you have "mapped" the Bam HI site.

8

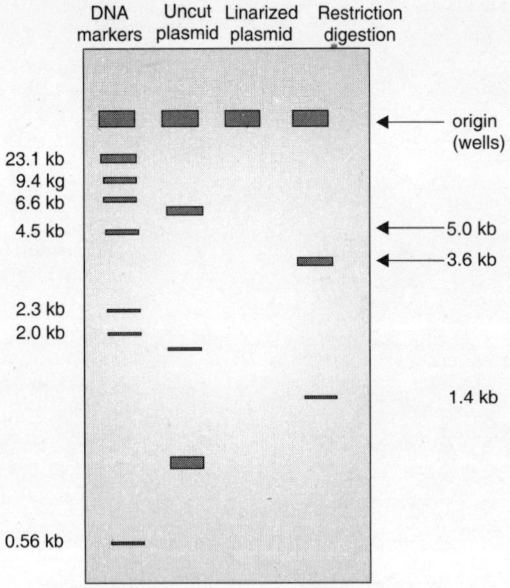

Fig. 8.4: Electrophoresis gel for above mentioned case

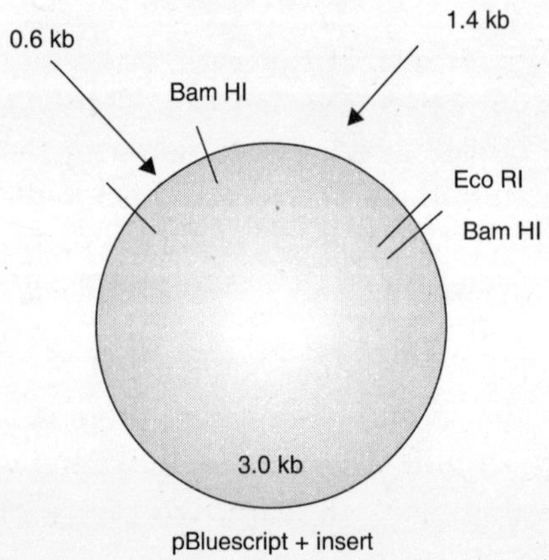

Fig. 8.5: Restriction map of pBluescript + insert

By testing the insert for the presence and location of sites of many different restriction enzymes, a "restriction map" of the clone is made. This will give us important structural information on the insert.

Uses of Restriction Mapping

Restriction map information is important for many techniques used to manipulate DNA. One application is to cut a large piece of DNA into smaller fragments to allow it to be sequenced. Genes and cDNAs can be thousands of kilobases long (megabases—Mb); however, they can only be sequenced 400 bases at a time. DNA must be chopped up into smaller pieces and subcloned to perform the sequencing. Also, restriction mapping is an easy way to compare DNA fragments without having any information of their nucleotide sequence. For example, you may isolate two clones for a gene that are 8 kb and 10 kb long. You know that they overlap, because the procedure you used to isolate them told you that they have sequences in common. A restriction map can tell you how much they overlap by.

From the restriction map information, you can tell which parts of the two clones are identical and which parts are different. The parts of the clones that overlap are identical. If you were interested in the sequence of this gene, you would only have to sequence the area of overlap in one of the clones, greatly reducing the amount of sequencing that you would have to do.

Procedure: The procedure for this experiment will be identical to the procedure of the last experiment. You will digest 3 µl of your plasmid with a restriction enzyme and then electrophorese it on an agarose gel. You will then determine if the insert has a site for that enzyme and where the site is based on the results of the agarose gel.

You will set up your restriction digestion like this:

14.5 µl H_2O

2.0 µl 10 Restriction enzyme buffer

3.0 µl plasmid DNA solution

0.5 µl restriction enzyme

20.0 µl total

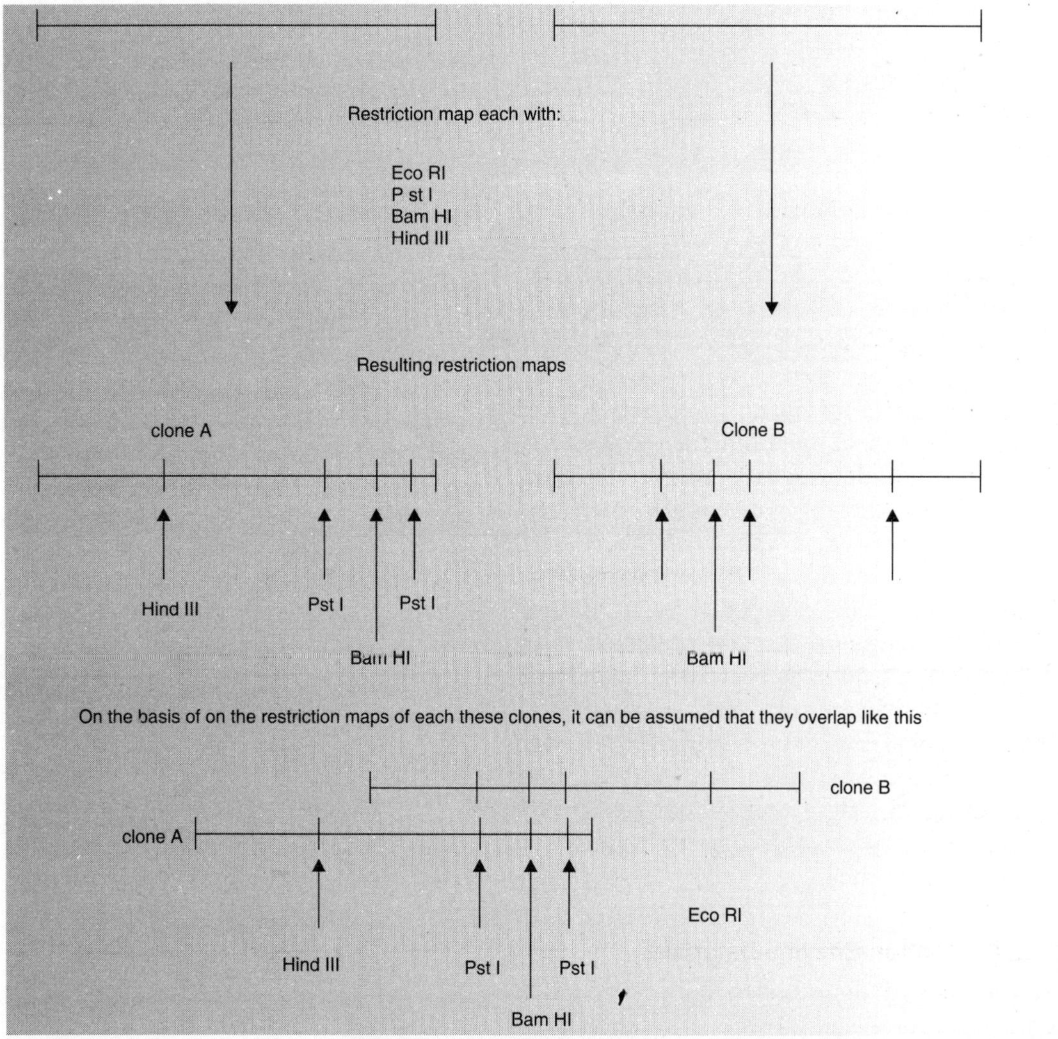

Fig. 8.6: Restriction mapping

When loading a gel, you should always load a DNA marker, your uncut plasmid and the vector plasmid (the plasmid without an insert) linearized. The DNA marker and linearized vector are used to estimate the sizes of your restriction fragments. They will be premeasured and provided on your lab bench. You will set up the above digestion and allow it to digest for 30 min. While this is occurring, you will put 18 µl of H_2O and 2 µl of your plasmid uncut in a tube. When the digestion is complete, add 2 µl of dye to each of the four tubes (marker, linearized vector, digested plasmid, and uncut plasmid) and then load them on an agarose gel.

Identifying Your Unknown

After the gel has finished running, stain it as before and take a picture. Determine the sizes of the fragments resulting from your restriction digests by graphing on semi-log paper and construct a restriction map of your clone from this data. Determine which insert you have in your plasmid from among the possible inserts.

REBASE

REBASE is a comprehensive database of information about restriction enzymes, DNA methyltransferases and related proteins involved in restriction-modification. It contains both published and unpublished work with information about recognition and cleavage sites, isoschizomers, commercial availability, crystal and sequence data. Experimentally characterized homing endonucleases are also included. Additionally, REBASE contains complete and up-to-date information about the methylation sensitivity of restriction endonucleases. An extensive analysis is included of the restriction-modification systems that are predicted to be present in the sequenced bacterial and archaeal genomes from GenBank. The contents of REBASE are available by browsing from the web (http://rebase.neb.com/rebase/rebase.html) and through selected compilations by ftp (ftp.neb.com) and as monthly updates that can be requested via e-mail.

The Restriction Enzyme Database

A collection of information about restriction enzymes and related proteins. It contains published and unpublished references, recognition and cleavage sites, isoschizomers, commercial availability, methylation sensitivity, crystal, genome, and sequence data. DNA methyltransferases, homing endonucleases, nicking enzymes, specificity subunits and control proteins are also included. Putative DNA methyltransferases and restriction enzymes, as predicted from analysis of genomic sequences, are also listed. REBASE is updated daily and is constantly expanding.

Link: http://rebase.neb.com/rebase/rebhelp.html

Primer Designing

A primer is a strand of nucleic acid that serves as a starting point for DNA replication. They are required because the enzymes that catalyze replication, DNA polymerases, can only add new nucleotides to an existing strand of DNA. The polymerase starts replication at the 3'-end of the primer, and copies the opposite strand. In most cases of natural DNA replication, the primer for DNA synthesis and replication is a short strand of RNA (which can be made de novo). This RNA is produced by primase, and is later removed and replaced with DNA by a repair polymerase. Many of the laboratory techniques of biochemistry and molecular biology that involve DNA polymerase, such as DNA sequencing and the polymerase chain reaction (PCR), require primers. These primers are usually short, chemically synthesized oligonucleotides, with a length of about twenty bases. They are hybridized to a target DNA, which is then copied by the polymerase (Fig. 8.7).

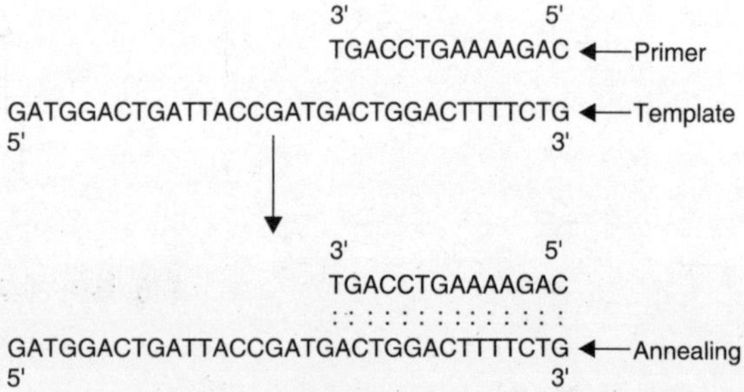

Fig. 8.7: Principle of primer design

Some thoughts on designing primers

1. Primers should be 17–28 bases in length;
2. Base composition should be 50–60% (G + C);
3. Primers should end (3') in a G or C, or CG or GC: this increases efficiency of priming;

4. Tms between 55-80 C are preferred;
5. 3'-ends of primers should not be complementary, as otherwise primer dimers will be synthesised preferentially to any other product;
6. Primer self-complementarity (ability to form 2 structures such as hairpins) should be avoided;
7. Runs of three or more Cs or Gs at the 3'-ends of primers may promote mispriming at G or C-rich sequences (because of stability of annealing), and should be avoided.

Also keep in mind that most oligonucleotide synthesis reactions are only 98% efficient. This means that each time a base is added, only 98% of the oligos will receive the base. This is not often critical with shorter oligos, but as length increases, so does the probability that a primer will be missing a base. This is very important in mutagenesis or cloning reactions. Purification by HPLC or PAGE is recommended in some cases.

Analysis of Primer Sequences using Primer 3

Primer 3 is a widely used program for designing PCR. Primer 3 can also design hybridization probes and sequencing primers. Primer 3 has many different input parameters that you control and that tell primer 3 exactly what characteristics make good primers for your goals.

DNA Strider

DNA Strider is a software program designed to automate and simplify analysis of DNA sequences. The Biology Dept at Lewis & Clark College has recently purchased a site license for version 1.3, which is fully compatible with Mac operating systems 7 through 9. Strider is designed to be a simple, affordable, straightforward program which is powerful enough to perform most sequence analyses without a lot of fancy extra features

DNA Strider is a new integrated DNA and protein sequence analysis program written with the C language. It has been designed as an easy to learn and use program as well as a fast and efficient tool for the day-to-day sequence analysis work. The program consists of a multi-window sequence editor and of various DNA and protein analysis functions. The editor may use 4 different types of sequences (DNA, degenerate DNA, RNA and one-letter coded protein) and can handle simultaneously 6 sequences of any type up to 32.5 kB each. Negative numbering of the bases is allowed for DNA sequences. All classical restriction and translation analysis functions are present and can be performed in any order on any open sequence or part of a sequence. The main feature of the program is that the same analysis function can be repeated several times on different sequences, thus generating multiple windows on the screen.

Many graphic capabilities have been incorporated such as graphic restriction map, hydrophobicity profile and the CAI plot-codon adaptation index. The restriction sites search uses a newly designed fast hexamer look-ahead algorithm. Typical runtime for the search of all sites with a library of 130 restriction endonucleases is 1 second per 10,000 bases. The circular graphic restriction map of the pBR322 plasmid can be therefore computed from its sequence and displayed on the Macintosh Plus screen within 2 seconds and its multiline restriction map obtained in a scrolling window within 5 seconds.

Link for Manual

http://www.lclark.edu/~bkbaxter/Bio312%20web/strider.htm

Link for downloading DNA Strider

http://mac.softpedia.com/get/Math-Scientific/DNAStrider.shtml

MacVector

MacVector is a comprehensive Macintosh application that provides sequence editing, primer design, internet database searching, protein analysis, sequence confirmation, multiple sequence alignment, phylogenetic reconstruction, coding region analysis, and a wide variety of other functions. MacVector is widely regarded as the most intuitive, easy to use program available for sequence analysis.

Link for MacVector

http://www.macvector.com/

The General Program Interface

MacVector adheres to the Macintosh Human Interface Guidelines, and provides an interface that is easy to use and familiar to any Macintosh user. The program menu bar has six items: File, Edit, Options, Analyze, Database, and Windows. The File, Edit, and Windows menus are similar to these menus in other Macintosh programs. The Options menu offers three commands that can be used to modify the way information is presented in analysis results windows (Format Annotated Display, Format Aligned Display, and Default Symbols). The user can modify or create the genetic code assignments to be used in all translations by selecting the Modify Genetic Codes command, or create a codon bias table for an organism by using the Make Codon Bias Table command in conjunction with the Entrez databases at the NCBI. The Analyze menu lists the various analysis functions that MacVector can perform on nucleic acid and protein sequences. The Database menu includes commands pertaining to database searching, such as BLAST searches, Entrez browsing and pairwise alignments of a query sequence to a user database of sequences saved locally.

The sequence is the central object in MacVector, and a sequence window must be active before an analysis can be performed on it. The program is highly interactive. The commands in the Analyze menu are available depending on what sequence type is active. When a nucleic acid sequence is active, all protein-analysis functions are grayed out, and vice versa. Most of the analyses store intermediate results in memory and allow the user to apply various filters to view different subsets of the results over and over without having to redo the entire analysis. Many analyses also offer a number of different ways in which to view the results: in tabular format, graphical format, and on the annotated sequence. The user can zoom in to view local regions of the graphical results in more detail. Results may be saved as text (tabular results) or PICT (graphical results) files, which may be reviewed or edited in word processing or drawing programs.

When a user initiates an analysis, a dialog box specific to that analysis appears wherein the user can either enter parameters of their choice or opt to perform the analysis with the default parameters. The user submits the analysis and once it is complete, MacVector presents the user with the results dialog box.

The results of most analyses in MacVector can be presented in more than one form, and the results dialog box presents the user with the choice of result formats, as well as any possible options to filter the results before they are presented. Once the choice of result formats is made, MacVector presents them.

Restriction Analyses

MacVector comes with the REBASE restriction enzyme database, which consists of several restriction enzyme files. The restriction enzyme files are lists of enzymes carried by different manufacturers. The File | Open command is used to open the restriction enzyme files: Each row has the name, cut site, and any known isoschizomers of a restriction enzyme. Clicking on an enzyme puts a check mark at the left of its name, and selects this enzyme for use in restriction analysis. Selecting the Analyze | Restriction Enzyme command brings up the restriction analysis dialog box. The user clicks on the Enzyme File button to select a restriction enzyme file for the analysis. The pull-down menu next to Search using offers the user the choice of using all enzymes in the restriction enzyme file or only the enzymes that were selected earlier.

Once the analysis is complete, the results dialog box lists the output options: a list of restriction enzymes that cut, a list of those that do not, the restriction map, the cut site annotated sequence, and finally the fragment prediction. The restriction map can be customized to include features and the results can be filtered so that only a subset of the cut sites is shown. The map can be toggled between circular and linear by clicking on the circular/linear button in the Graphics Palette. Fragments can be predicted using single or double digests.

Major Features

- Graphical sequence editing
- Primer design
- DNA analysis
- Protein analysis
- Database searching
- Multiple sequence analysis
- Sequence assembly
- Primer design

Primer Analysis

MacVector can predict PCR primer pairs for a given template, based on parameters selected by the user. Here the user provides MacVector with criteria for an acceptable primer, the program scans the sequence and presents the user with a list of primers, after eliminating those primers that do not match the user's criteria. MacVector can also test primers against a template sequence for usability in PCR. In this case, the program analyzes the primer input by the user, and provides information about the primer. The user can then decide, based on the results of the analysis, whether the primer should be used in an experiment. Similarly, MacVector can also predict primers to be used in sequencing a template and test a user-defined primer against a template for its usability as a sequencing primer. Whereas the PCR primer-analysis functions predict or test primer pairs, the sequencing primer-analysis functions predict or test single primers.

PCR primer-pair prediction is initiated by the Analyze | Primers | PCR Primer Pairs command; sequencing primer prediction by Analyze | Primers | Sequencing Primers Probes. The initial dialog allows the user to specify the region of interest; for PCR primer pairs, it is based on either product size or flanking regions. Other parameters that are user-specified include range of melting temperature for the primer-template duplex, G+C content of the primers, range of desired primer length, and whether a G-C anchor is required in the primer. To eliminate primers with too much secondary structure, the user specifies the maximum number of consecutive bonds the primer is allowed to have when it forms a hairpin with itself or a duplex with another primer. Primers that bind to other sites on the sequence are also eliminated. In the case of PCR primer pairs, the program compares the 3′ end of each primer with the product that will be amplified in the reaction. For sequencing primers, the 3′ end of each primer is compared with the entire sequence, or a specified region of the sequence.

The results dialog box presents a list of statistics, including the number of primers tested, number of primers accepted and reasons why primers were rejected, and the choice of viewing the results as a list or a graphical map. The list of PCR primer pairs provides Tm and %GC information for each primer of a pair, the length of the product, the product's Tm and %GC, and the optimum temperature to be used for the reaction.

User-defined primer testing is initiated by the command Analyze | Primers | Test PCR Primer Pairs (or Analyze | Primers | Test Sequencing Primers/Probes). A dialog box appears, in which one or two primers can be entered (for testing sequencing primers there is space for only one primer). Once the user clicks on the Apply button, statistics for the primers appear Information provided includes Tm and %GC for the primers, whether the primers actually anneal to the template, and whether the primers show any excessive secondary structure. In the case of user-defined PCR primer pairs, additional information pertinent to PCR is provided, including the size of the product formed, if any, whether the 3′ end of either primer binds within the product and the Tm difference of the primer pair. Any information that would be detrimental to the success of the reaction is highlighted in red.

Hybridization probes for a nucleic acid sequence can be predicted in the same fashion as sequencing primers. To obtain a hybridization probe for a gene whose exact sequence is not known, but whose amino acid sequence is known, the reverse translation function can be used. MacVector then scans the resulting DNA sequence to find a region that is not too degenerate and outputs a list of oligos along with their dissociation temperatures, G+C content and the number of oligos that would have to be synthesized to cover all possible sequences represented by the degenerate oligo. The user can modify the genetic code used in the reverse-translation by reducing the number of codons each amino acid has (based on knowledge of the organism's codon bias). This will reduce the degeneracy in the hybridization probes predicted by MacVector.

Vector NTI

Vector NTI is a bioinformatics software package. It is available for free to academic, governmental, and non-profit researchers. Current versions are v10 for Windows/PCs and v7.1 for Macs but only Mac OS X v10.3 (Panther) not Mac OS X v10.4 (Tiger). It is a software package to aid genetic engineers, molecular biologists, students and others in the study, manipulation and creation of biological molecules. It is the most highly integrated, multifunctional sequence analysis. It contains a comprehensive set of data analysis and management tools.

It has five modules Vector NTI sequence creation, mapping, analysis, annotation, illustration and management.

Features

- Create, annotate, analyse, and share DNA/protein sequences.
- Perform and save BLAST searches.
- Design primers for PCR, cloning, sequencing or hybridisation experiments.
- Plan cloning and run gels in silico.
- Align multiple protein or DNA sequences.
- Search NCBI's Entrez, view, and save DNAs, proteins, and citations.
- Edit chromatogram data, assemble into contigs .
- Molecular viewer.

Omiga

The Omiga DNA and Protein Sequence Analysis Software tool, provides an effective and comprehensive tool for the analysis of both nucleic acid and protein sequences and runs on the ubiquitous standard PC. Omiga allows the import of sequences in several common formats. Upon importing sequences and assigning them to various projects, Omiga allows the user to produce, analyze, and edit sequence alignments. Sequences may also be queried for the presence of restriction sites, sequence motifs, and other sequence features, all of which can be added into the notations accompanying each sequence. Finally, Omiga allows rapid searches for putative coding regions as well as PCR and sequencing primers.

Omiga allows the user to identify restriction sites and proteolytic sites in nucleotide and peptide sequences, respectively. When searching for restriction endonuclease cleavage sites, as with searches for proteolytic sites, the user may search for each type of site in turn by inputting the actual sequence recognized by the endonuclease or protease in question. Alternatively, Omiga contains the REBASE database of all common restriction endonuclease sites as well as a PABASE database of proteolytic sites that can be used to identify all sites within a sequence.

This is done by selecting either the restriction sites or proteolytic sites option from the pull-down menu generated by selecting the search feature from the tool bar. The output thus generated and displayed in the search results view can then be filtered to identify sites ideal for subcloning reactions. In this way, Omiga may serve as a valuable tool for planning experiments. An example of this is provided in the tutorials that are included with Omiga. In these tutorials, the novice user is led through a process of importing the human β-globin domain sequence, and identifying restriction endonuclease sites that will allow subcloning of the β-globin gene into the vector pUC 19 in a single ligation. The new pUC–β-globin construct can then be saved as a new sequence and searched for restriction endonuclease sites that will give a characteristic restriction pattern to verify the success of the ligation and subsequent cloning reactions in the laboratory. Especially with large sequences, such as the approx 75 kb human β-globin domain (GenBank locus HUMHBB), this is far preferable to identifying such sites simply by inspection with the unaided eye.

Primer Identification

Both sequencing and PCR primers can be designed using user-defined criteria. This is carried out by selecting a sequence in the tree panel, then selecting PCR Primer Pairs or Sequencing Primers from the Search pull-down menu. Default protocols are provided, however, the user may also define search parameters of his or her own and has the option of saving these as additional protocols. When searching for PCR primers, the parameters over which the user has control include primer length, GC content, melting temperature, salt concentration, and primer concentration. The user may specify specific regions of the template sequence to be searched or omitted from the search.

The user may specify a 3′ clamp, as well as a number of ambiguous nucleotides for degenerate primers. Another feature available for primer design is the ability to omit duplicate end points. That is, if one 5′ primer is compatible with numerous potential 3′ primers, the user has the option to show only that compatible 3′ primer that produces the shortest amplification product, thus reducing the clutter of acceptable pairs that can result from identifying primer pairs in a very large template sequence. However, there does not seem to be a feature that allows the user to design primers compatible with an already defined primer. For example, it may be desirable to design a primer pair that will amplify across an intron (to differentiate genomic from cDNA), and then design an additional 3′ primer that is compatible with the 5′ primer of the original pair. The additional 3′ primer could be designed within an intron, to identify premRNA, or within a different exon to show alternative splicing. The ability to identify new primers compatible with a preselected primer for PCR amplification is a common and useful feature of several other primer-design tools.

When identifying PCR primer pairs, Omiga will identify three types of primer-to-primer interactions, including primer–primer annealing, primer self-annealing, and primer – template annealing. The way in which Omiga handles these three types of annealing can be specified by the user. Primer–primer interactions include the dreaded primer—dimer formation. If very stable primer–dimers are formed, particularly at the 3′ end of the primers, little or no amplification of template will occur. Primer self-annealing deals with hairpin formation within each individual primer. As with primer–dimers, if too large a fraction of an individual primer is in the form of hairpin structures, little or no amplification will occur, as much of the one primer will be sequestered in the hairpin form, and will be unavailable for annealing to the template. Finally, Omiga will also search for nonspecific primer annealing within a template sequence. This is important to avoid primers that contain low-complexity sequence elements, and those primers that may be directed towards repetitive regions within the template. An additional feature that is missing from Omiga's primer design capabilities is the ability to search primers against additional sequence databases. Whereas it can often be time consuming to compare primer sequences against large databases, it is also often a very valuable functionality. For example, when designing primers for the amplification of regions within the genome of a very simple organism for which the entire genome is known, it might be desirable to compare your primers to the entire genome of that organism, to avoid spurious priming. In the age of whole genomes, such a search will become increasingly possible. Similarly, if primers are designed within a short sequence from the public databases, it may be desirable to scan them against a database of known repetitive elements. Whereas such sequences are rare within coding sequence, and thus not likely to be identified by the primer template annealing comparisons performed by Omiga, the presence of even a slight similarity to an Alu or Line elements will result in frequent inappropriate priming and poor amplification of the desired amplicon.

9

Molecular Visualization

The main feature of computer visualization programs is interactivity, which allows users to visually manipulate the structural images through a graphical user interface. At the touch of a mouse button, a user can move, rotate, and zoom an atomic model on a computer screen in real-time, or examine any portion of the structure in great detail, as well as draw it in various forms in different colors. Further manipulations can include changing the conformation of a structure by protein modelling or matching a legend to an enzyme active site through docking exercises.

Because a Protein Data Bank (PDB) data file for a protein structure contains only x, y, and z coordinates of atoms, the most basic requirement for a visualization program is to build connectivity between atoms to make a view of a molecule. The visualization program should also be able to produce molecular structures in different styles, which include wire frames, balls and sticks, space-filling spheres, and ribbons.

A wire-frame diagram is a line drawing representing bonds between atoms. The wire frame is the simplest form of model representation and is useful for localizing positions of specific residues in a protein structure, or for displaying a skeletal form of a structure when Cα atoms of each residue are connected. Balls and sticks are solid spheres and rods, representing atoms and bonds, respectively. These diagrams can also be used to represent the backbone of a structure. In a space-filling representation (or Corey, Pauling, and Koltan [CPK]), each atom is described using large solid spheres with radii corresponding to the van der Waals radii of the atoms.

Ribbon diagrams use cylinders or spiral ribbons to represent α-helices and broad, flat arrows to represent β-strands. This type of representation is very attractive in that it allows easy identification of secondary structure elements and gives a clear view of the overall topology of the structure. The resulting images are also visually appealing.

Different representation styles can be used in combination to highlight a certain feature of a structure while de-emphasizing the structures surrounding it. For example, a cofactor of an enzyme can be shown as space-filling spheres while the rest of the protein structure is shown as wireframes or ribbons (Fig. 9.1). Some widely used and freely available software programs for molecular graphics are introduced in this chapter.

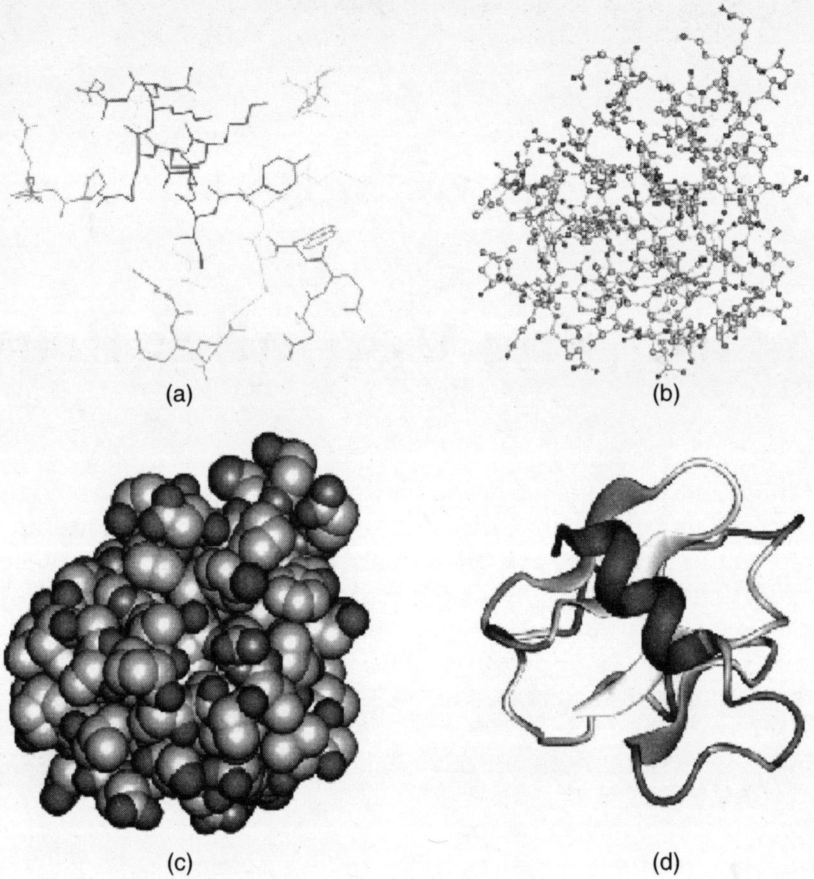

Fig. 9.1: Examples of molecular structure visualization forms–(a) Wireframes, (b) Balls and sticks, (c) Space-filling spheres, (d) Ribbons

Usages of Visualization Software Available in Public Domain

VMD

VMD (http://www.ks.uiuc.edu/Research/vmd/) is a molecular visualization program for displaying, animating, and analyzing large biomolecular systems using 3D graphics and built-in scripting. VMD supports computers running MacOS-X, Unix, or Windows, is distributed free of charge, and includes source code.

VMD is designed for modelling, visualization, and analysis of biological systems such as proteins, nucleic acids, lipid bilayer assemblies, etc. It may be used to view more general molecules, as VMD can read standard Protein Data Bank (PDB) files and display the contained structure. VMD provides a wide variety of methods for rendering and coloring a molecule—simple points and lines, CPK spheres and cylinders, licorice bonds, backbone tubes and ribbons, cartoon drawings, and others. VMD can be used to animate and analyze the trajectory of a molecular dynamics (MD) simulation. In particular, VMD can act as a graphical front end for an external MD program by displaying and animating a molecule undergoing simulation on a remote computer.

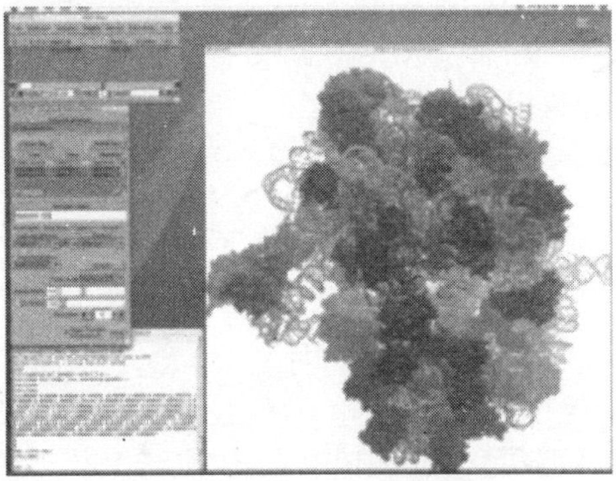

Fig. 9.2: VMD screen shot

Key features of VMD

- Support for all major computer platforms
- Support for multicore processors
- Support for GPU accelerated computation
- Many excellent VMD tutorials developed locally, and by the research community at large
- No limits on the number of molecules, atoms, residues or number of trajectory frames, except available memory
- Many molecular rendering and coloring methods
- Stereo display capability
- Extensive atom selection syntax for choosing subsets of atoms for display (includes boolean operators, regular expressions, and more)
- Support for over 60 molecular file formats and data types through an extensive library of built-in file reader/writer plugins and translators
- VMD includes a multiple sequence alignment plugin, a unified bioinformatics analysis environment that allows one to organize, display, and analyze both sequence and structure data for proteins and nucleic acids.
- Ability to export displayed graphics to files which may be processed by a number of popular ray tracing and image rendering packages, including POV-Ray, Rayshade, Raster3D, and Tachyon.
- User-extensible graphical and text-based user interfaces, built-on standard Tcl/Tk and Python scripting languages
- Extensions to the Tcl language which enable researchers to write their own routines for molecular analysis
- Modular, extensible source code using an object-oriented design in C++, with a programmers guide describing the program architecture and source code
- Integration with the program NAMD, a fast, parallel, and scalable molecular dynamics program developed in conjunction with VMD in the Theoretical and Computational Biophysics Group at the University of Illinois. See the NAMD WWW home page for more info: http://www.ks.uiuc.edu/Research/namd/
- VMD works well with projected display systems like the 3D Projection Facility maintained by the Theoretical and Computational Biophysics Group.
- VMD can be used to concurrently display and interact with a running NAMD simulation.

Ras Mol

RasMol (http://www.umass.edu/microbio/rasmol/) is a computer program written for molecular graphics visualization intended and used primarily for the depiction and exploration of biological macromolecule structures, such as those found in the Protein Data Bank. It was originally developed by Roger Sayle in the early 90s.

Historically, it was an important tool for molecular biologists since the extremely optimized program allowed the software to run on (then) modestly powerful personal computers. Before RasMol, visualization software ran on graphics workstations that, due to their expense, were less accessible to scholars. RasMol has become an important educational tool as well as continuing to be an important tool for research in structural biology.

RasMol has a complex version history. Starting with the series of 2.7 versions, RasMol is licensed under a dual license (GPL or custom license RASLIC). Thus, RasMol is (along with Molekel, Jmol and PyMOL), among the few open source molecular visualization programs available.

RasMol includes a language (for selecting certain protein chains, or changing colors, etc). Jmol has incorporated the RasMol scripting language into its commands.

Protein Data Bank (PDB) files can be downloaded for visualization from the Research Collaboratory for Structural Bioinformatics (RCSB) bank. These have been uploaded by researchers who have characterized the structure of molecules usually by X-ray crystallography or NMR spectroscopy.

Pymol

PyMOL (http://www.pymol.org/) is an open-source, user-sponsored, molecular visualization system created by Warren Lyford DeLano and commercialized by DeLano Scientific LLC, which is a private software company dedicated to creating useful tools that become universally accessible to scientific and educational communities. It is well suited to producing high quality 3D images of small molecules and biological macromolecules such as proteins. According to the author, almost a quarter of all published images of 3D protein structures in the scientific literature were made using PyMOL.

PyMOL is good for:

1. Viewing 3D structures
2. Rendering figures
3. Giving presentations
4. Animating molecules
5. Sharing visualizations
6. Exporting geometry

PyMOL is one of few open source visualization tools available for use in structural biology. The Py portion of the software's name refers to the fact that it extends, and is extensible by the Python programming language.

SPDB Viewer

(http://expasy.org/spdbv/)

Swiss-PdbViewer (or SPDBV), is an interactive molecular graphics program for viewing and analyzing protein and nucleic acid structures. In combination with Swiss-Model (a server for automated comparative protein modelling maintained at http://www.expasy.org/swissmod) new protein structures can also be modelled.

A molecular coordinate file (e.g. *.pdb, *.mm CIF, etc.) is a text file containing, amongst other information, the atom coordinates of one or several molecules. It can be opened from a local directory or imported from a remote server by entering its PDB accession code. The content of one coordinate file is loaded in one (or more) layers, the first one will be referred to as the "reference layer".

DeepView can simultaneously display several layers, and this constitutes a project. When working on projects, the layer that is currently governed by the Control Panel is called the currently active layer. Each molecule is composed of groups, which can be amino acids, hetero-groups, water molecules, etc. and each group is composed of atoms.

Non-coordinate files containing specific information other than atom coordinates. Molecular surfaces, electrostatic potential maps, and electron density maps are examples of non-coordinate files, which can either be computed by DeepView, or loaded from specialized external programs.

CHIME

MDL Chime (http://www.umass.edu/microbio/chime/getchime.htm) is a free plugin used by web browsers to display the three-dimensional structures of molecules. It is part of the ISIS product line acquired by Symyx Technologies from scientific publisher Elsevier in October 2007. It is based on the RasMol code.

Chime is used by a wide range of biochemistry web sites for the visualization of macromolecules. Many of these sites are linked to the World Index of Molecular Visualization Resources MolVisIndex.Org. Chime was also used until 2006 at the Protein Data Bank to examine structures stored there.

Although available in 1998 in both Windows 98 and Mac OS 9 versions for both Netscape and Internet Explorer browsers, development of Chime did not follow the move to Mac OS X for the Mac and support for Windows-based browsers other than Internet Explorer was dropped (although it works well in Mozilla Firefox). No significant features have been added since 1998 except for updates in the installation package to follow the development of Windows and Internet Explorer.

Chime may be superseded by Jmol, but the current market share and trends are unclear. Jmol is a non-proprietary Java molecular visualization application/applet that has maintained most Chime command compatibility while adding features.

Two features of Chime which are not yet reproduced with Jmol:

- *'Sculpt' mode:* The ability to change the conformation of molecules in a realistic way by dragging their atoms to new positions. This is a useful feature for exploring the conformational possibilities of a molecule such as the inability to rotate a carbon-carbon double bond.
- Calculating and drawing molecular surfaces coloured according to electrostatic or hydrophobic potential. Chime can calculate these surfaces for any molecule.

Cn3D

Cn3D (http://www.ncbi.nlm.nih.gov/Structure/CN3D/cn3dtut.shtml) is a Windows, Macintosh and Unix-based software from the United States National Library of Medicine that acts as a helper application for web browsers to view three-dimensional structures from The National Center for Biotechnology Information's Entrez retrieval service. It "simultaneously displays structure, sequence, and alignment, and now has powerful annotation and alignment editing features", according to its official site.

Cn3D is a helper application for your web browser that allows you to view three-dimensional structures from NCBI's Entrez retrieval service. Cn3D runs on Windows, Macintosh, and Unix. Cn3D simultaneously displays structure, sequence, and alignment, and now has powerful annotation and alignment editing features.

New features in Cn3D 4.1

- Improved sequence alignment algorithms.
- Automatic VAST alignment import.
- Taxonomy viewer.

New features in Cn3D 4.0

- Re-written in C++ to improve performance and extendibility.
- Improved OpenGL rendering speed.
- New user interface using wxWindows.
- Complete alignment editing system, along with algorithms for constructing new alignments.
- Extensive structure and alignment annotation features.
- Detailed style control.
- High-resolution image export.
- Built-in help system.

What does Cn3D do?

Cn3D is a visualization tool for biomolecular structures, sequences, and sequence alignments. What sets Cn3D apart from other software is its ability to correlate structure and sequence information, e.g. a scientist can quickly find the residues in a crystal structure that correspond to known disease mutations, or conserved active site residues from a family of sequence homologs. Cn3D displays structure-structure alignments along with their structure-based sequence alignments, to emphasize what regions of a group of related proteins are most conserved in structure and sequence. Also included are custom labelling features, high-quality OpenGL graphics, and a variety of file exports that together make Cn3D a powerful tool for literature annotation. Cn3D is typically run from a WWW browser as a helper application for NCBI's Entrez system, but it can also be used as a stand alone application.

With version 4, Cn3D is now a complete multiple alignment editor as well, and includes algorithms for aligning sequences to other sequences and to structures. You can now create and even annotate multiple alignments. Cn3D is used as the primary alignment curation tool for the CDD project.

Cn3D is available for Windows, Macintosh, and various Unix platforms.

Other 3D viewers: Mage, CAD, and VRML

A variety of file formats have been used to present three-dimensional biomolecular structure data lacking in chemistry-specific data representations. These are viewed in generic three-dimensional data viewers such as those used for "macroscopic" data, like engineering software or virtual-reality browsers. File formats such as VRML contain three-dimensional graphical display information but little or no information about the underlying chemical graph of a molecule. Furthermore, it is difficult to encode the variety of rendering styles in such a file; one needs a separate VRML file for a space-filling model of a molecule, a wire-frame model, a ball-and-stick model, and so on, because each explicit list of graphics objects (cylinders, lines, spheres) must be contained in the file.

Biomolecular three-dimensional structure database records are currently not compatible with "macroscopic" software tools such as those based on CAD software. Computer-aided design software represents a mature, robust technology, generally superior to the available molecular structure software. However, CAD software and file formats in general are ill-suited to examine the molecular world, owing to the lack of certain "specialty" views and analytical functions built in for the examination of details of protein structures.

Molscript

Molscrip (www.avatar.se/molscript/) is a Unix program capable of generating wireframe, space-filling, or ball-and-stick styles. In particular, secondary structure elements can be drawn with solid spirals and arrows representing α-helices and β-strands, respectively. Visually appealing images can be generated that are of publication quality. The drawback is that the program is command-line-based and not very user-friendly. A modified Unix program called Bobscript (www.strubi.ox. ac.uk/bobscript/) is available with enhanced features.

WebMol

Webmol (www.cmpharm.ucsf.edu/cgi-bin/web mol.pl) is a web-based program built based on a modified RasMol code and thus shares many similarities with RasMol. It runs directly on a browser of any type as an applet and is able to display simple line drawing models of protein structures. It also has a feature of interactively displaying Ramachandran plots for structure model evaluation.

10

In Silico Modelling

In silico modelling can be considered as a range of computerized techniques based on theoretical chemistry methods and experimental data that can be used either to analyze molecules and molecular systems or to predict molecular, chemical, and biochemical properties. It can be defined as the generation, manipulation, calculation, and prediction of realistic molecular structures and associated physicochemical as well as biochemical properties by the use of a computer. It is primarily a mean of communication between scientist and computer, the imperative interface between human-comprehensive symbolism, and the mathematical description of the molecule.

Applications of In Silico Modelling

It serves as a bridge between theory and experiment to:

1. Extract results for a particular model.
2. Compare experimental results of the system.
3. Compare theoretical predictions for the model.
4. Help understanding and interpreting experimental observations.
5. Correlate between microscopic details at atomic and molecular level and macroscopic properties.
6. Provide information not available from real experiments.
7. Explore potential drug target and design of suitable inhibitors.

In silico Modelling methods are now routinely used to investigate the structure, dynamics and thermodynamics of inorganic, biological, and polymeric systems. The types of biological activity that have been investigated using molecular modelling include protein folding, enzyme catalysis, protein stability, conformational changes associated with biomolecular function, and molecular recognition of proteins, DNA, and membrane complexes.

Popular software for In silico Modelling

- Abalone
- AMBER
- ADF
- Ascalaph designer
- BALL view
- Biskit
- BOSS
- Cerius2
- CP2K
- Chimera
- CHARMM
- Coot
- COSMOS

- CP_2K
- CPMD
- Culgi
- Firefly
- GAMESS
- GAUSSIAN
- Ghemical
- GROMACS
- GROMOS
- InsightII
- LigandScout
- LAMMPS
- MacroModel
- MarvinSpace
- Materials Studio
- MDynaMix
- MMTK
- Molecular operating environment (MOE)
- Molecular Docking Server
- MOPAC
- NAMD
- NOCH
- Oscail X
- PyMOL
- Q-Chem
- Sirius
- SPARTAN
- TURBOMOLE
- ReaxFF
- VMD
- WHAT IF
- xeo
- YASARA
- Zodiac (software)

Thus functions of In silico modelling include:

- *Structure retrieval or generation:* Crystal structures of organic compounds can be found in the Cambridge Crystallographic Datafiles (http://www.ccdc.cam.ac.uk/). Those that do not exist may be generated by 3D rendering software. The 3D structural coordinates of biomacromolecules can be retrieved from Protein Data Bank (http://www.rcsb.org/pdb/).
- *Structural visualization:* Computer graphics is the most effective means for visualization and interactive manipulation of molecules and molecular systems. Numerous software programs (e.g. Cn3D, RasMol and KineMage) are available for visualization, management, and manipulation of molecular structures.
- *Energy calculation and minimization:* One of the fundamental properties of molecules is their energy content and energy level. Three major theoretical computational methods of their calculation include empirical (molecular mechanics), semiempirical, and ab initio (quantum mechanics) approaches. Energy minimization results in geometry optimization of the molecular structure.
- *Dynamics simulation and conformation search:* Solving motion of nuclei in the average field of the electrons is called quantum dynamics. Solution to the Newton's equation of motion for the nuclei is known as molecular dynamics. Integration of Newton's equation of motion for all atoms in the system generates molecular trajectories. Conformation search is carried out by repeating the process by rotating reference bonds (dihedral angles) of the molecule under investigation for finding lowest energy conformations of molecular systems.
- *Calculation of molecular properties:* Methods of estimating or computing properties (i.e. interpolating properties, extrapolating properties, and computing properties). Some computing properties are boiling point, molar volume, solubility, heat capacity, density, thermodynamic quantities, molar refractivity, magnetic susceptibility, dipole moment, partial

atomic charge, ionization potential, electrostatic potential, van der Waals surface area, and solvent accessible surface area.

- *Structure superposition and alignment:* Computing activities and properties of molecules often involves comparisons across a homologous series. Such techniques require superposition or alignment of structures.
- *Molecular interactions, docking:* The intermolecular interaction in a ligand-receptor complex is important and requires difficult Modelling exercises. Usually the receptor (e.g. protein) is kept rigid or partially rigid while the conformation of ligand molecule is allowed to change.

Comparative Modelling

Proteins from different sources and sometimes diverse biological functions can have similar sequences, and it is generally accepted that high sequence similarity is reflected by distinct structure similarity. Indeed, the root mean square deviation (RMSD) of the alpha-carbon coordinates for protein cores sharing 50% residue identity is expected to be around 1A°. Thus the most reliable prediction methods, termed comparative protein Modelling (also often called Modelling by homology), consist of the extrapolation of the structure of related family members (templates).

Identification of Modelling Templates

Comparitive protein Modelling requires at least one sequence of known 3D structure with significant similarity to the target sequence, In order to determine if a Modelling request can be carried out, one generally compares the target sequence with a database of sequences derived from the Brookhaven Protein Data Bank (PDB). This can be performed using any sequence similarity search algorithm. For a given target sequence, this procedure often allows the selection of more than one suitable Modelling template. In some cases, although this is by far not a general rule, multiple templates can help improve the resulting models. In our hands, however, there is no statistically relevant correlations between the number if templates used during model building and the over all quality of a model (as assessed during the 3D crunch experiment). In this case, the template structure sharing the highest sequence similarity with the target should serve as the reference on to which the other selected templates must be optimally superposed (in 3D). As a result, each residue of the reference structure is then aligned with a residue from the other available templates. This yields a structurally corrected multiple sequence alignment.

Constructing an Initial Model

The target sequence can then be aligned with the template sequence or, if several templates were selected, with the structurally corrected multiple sequence alignment using the best-scoring diagonals obtained by sequence alignment algorithms such as SIM. Residues, which should not be used for model building, for example those located in non-conserved loops, will be ignored during the Modelling process. Thus, the common core of the target protein and the loops completely defined by at least one supplied template structure will be built.

The sequence alignment guiding the Modelling procedure is a crucial component of the overall process and the quality of resulting models depends on it to a large extend. Therefore a high level of attention should be given to this step. It is generally accepted that the use of additional sequences (without known 3D structure) of the same protein family, selected to represent the sequence diversity within that family, can markedly improve the multiple sequence alignment. Such additional sequences can obviously be found by using the BLAST and their relative alignment can be found by searching the database such as SWISS-PROT and trEMBL using BLAST and their relative alignment can be found in protein family databases such as PFAM. Furthermore, it should be noted that manual adjustments are often the most powerful way to create corrected alignments where no algorithm is available.

Building the Model

The co-ordinates of the model can then be built. The (multiple) sequence alignment serves as a 'correspondence table' between target sequence and the template structures from which a weight averaged structural framework can be derived. This framework can then be completed by the addition of lacking or incomplete loop structures—which correspond to insertion and deletions in the sequence alignment—the rebuilding of undefined backbone atoms, the correction of ill-defined side chain geometries and the addition of lacking side chains.

Model Refinement and Manipulations

The final step of the co-ordinate generation process in the idealization of stereochemistry of the model, and consists mainly of the optimization of bond geometry and the removal of unfavorable non-bonded contacts. This step can be performed with energy minimization packages such as CHARM or GROMOS. Excessive energy minimization will cause the model to deviate markedly from the original model, which is not suitable and should be avoided. Indeed, experience has shown that changes induced by force field computations do not improve the accuracy of the model with respect to a control experimental structure. Thus, one should keep the number of minimization cycles to a minimum, but sufficient to improve the stereochemistry of the model.

A good protein structure should:

1. Minimize the number of torsion angles in disallowed regions of the Ramachandran plot
2. Maximize the number of H-bonds
3. Minimize the number of hydrophobic residues
4. Maximize the number of polar or charged residues
5. Minimize the number of non-bonded atoms within 2.6Å
6. Minimize the standard deviation in H-bonds energies
7. Minimize the standard deviation in dihedral angles for helices
8. Have low R-factor (<0.20 Å for X-ray structure) or a low backbone RMSD (0.8Å for NMR structures).

R-factor: It is a measure of the agreement between the calculated structure and the experimental data – for good protein structure it is 0.25, whereas for small molecule it is 0.05.

Molecular Superposition and Structural Alignment

- Like sequence comparison, structure comparison is also a prime objective of structural bioinformatics. But structure comparison is a much more computationally difficult process than sequence comparison.
- In sequence comparison, it is possible to use dynamic programming methods to generate alignment easily and rapidly used to identify regions of sequence similarity. But in structure comparison we compare the detailed 3D structure of molecule, it is quite a tedious task.
- There are certain computational tools and technique that are used to compare identical or relatively similar 3D structures. The most common method is called structure superposition. Superpo-

sition or superimposition is simply the process of rotating or orienting a structure on top of another structure until it becomes superimposed.

- The simplest method for 3D superimposition is to identify a minimum of 2 sets of common reference points; one set for the object to be superimposed and another set on the reference object that is being overlaid. As soon as these points are identified, the object to be superimposed can be rotated and translated until the 2 sets of reference points are almost matching.
- Certain mathematical approaches are available for superposition of molecules. These include 'Lagrangian multipliers', 'Quaternion' method and 'Matrix diagnolization'.
- It is possible to use this method to superimpose more than 2 structures. In this case, the 2 structures are superimposed to create a single structure which is then used as a template for the superimposition of the third structure. This process is repeated until all the structures have been superimposed. Generally, the two most similar structure are superimposed first, the least similar structure being superimposed last like an MSA.

Tools for molecular superimposition

- Two protein visualization programs are good for performing super-imposition. They are:
  1. Deep view
  2. MolMol
- There are two web servers that allows user to superimpose two structures (either through PDB IDs, SCOP IDs or through uploaded files). They are Superpose and Prosup.
- These servers use initial sequence alignment to help identify reference points so that superposition can be carried out against proteins with different number of amino acids. The output for these web servers include PDB co-ordinates list, which can be viewed by any number of visualization tools, images, different distance matrices, information on the alignment, number of equivalent residues and root mean square deviation (RMSD).
- Protein sequence alignment is evaluated either in terms of an Expect value (E value), a bit score or percent identity. In the case of structure comparison or structure alignment are evaluated by calculating RMSD (in Å).
- RMSD is a term for standard deviation i.e. RMSD is calculated the same way as standard deviation is calculated.
- After two structures are superimposed, the sum of squares of the distances between Cα atoms or equivalent C atom is calculated and divided by the number of atoms compared. The square root of this number is the RMSD.
- When more than two structures are superimposed, an average structure is created and whose RMSD is calculated first. The sum of the distance differences are then calculated relative to this hypothetical average structures.
- RMSD values are used frequently by NMR spectroscopists, structure modelers and X-ray crystallographers when comparing structures or characterizing structure families.
- If the RMSD value is >12, the structure belongs to random coil and completely unrelated.
- If the RMSD value is around 7.0Å, the structure has major problems and poor relationship similarity.
- If the structures have 1.5 Å RMSD, those structures are closely related and barely acceptable NMR structures.

- Finally if the RMSD value of structures is 0.8 Å, such structures are typical NMR structures and good structurally related,
- RMSD value between 0.4 and 0.8 Å indicates best NMR structures and good quality structures.

RAMS: (root mean square deviation): An archaic term for standard deviation. RMSD is used in the quantification of atomic position differences between protein structures. Very similar structures have RMSD values between 0 and 1.5Å; moderately similar structures have RMSD value between 1.5 and 3.0 Å.

The other computational tools for molecular superposition includes DISC (distance comparisons), which generate low energy conformation of molecules by a programme called ALADDIN.

The other methods for molecular superposition are:

1. 3D QSAR (quantitative structure activity relationship)
2. Ensemble distance geometry method.

Assessing the Model Quality

The quality of the model is determined by two distinct criteria, which will determine its applicability. First, the correctness of a model is dictated by the quality of the sequence alignment used to guide the Modelling process. If the sequence alignment is wrong in some regions, then the spatial arrangement of the residues in this portion of the model will be incorrect. Second, the accuracy of a model is essentially limited by the deviation of the used template structure(s) relative to the experimental control structure. This limitation is inherent to the methods used, since the models result from an extrapolation. As a consequence, the core Cα atoms of the protein models which share 35–50% sequence identity with their templates, will generally deviate by 1.0–1.5 A° from their experimental counter-parts as do experimentally elucidated structures. One should, however, not overlook the contributions of the templates to the model accuracy. The templates, which are obtained through experimental approaches, are subject to structural variations not only caused by experimental errors and differences in data collection conditions—such as the temperature, but also because of the different crystal lattice contacts and the presence or absence of the ligands. Almost every protein of model based on templates with a sequence similarity below 40% contain non-conserved loops, which are expected to be the least reliable portions of a protein model. Indeed, non-conserved loops generally deviate markedly from the experimentally determined control structures. In many cases, however, these loops also correspond to the most flexible parts of the structure as evidenced by their high crystallographic temperature factors (or multiple solutions in NMR experiments). On the other hand, the core residues—the least variable in any given protein family are usually found in essentially the same orientation as in experimental control structures, while far larger deviations are observed for surface amino acids.

This is expected since the core residues are generally well conserved and the rotamers of their side chains are constrained by neigbouring residues. In contrast, the more variable surface amino acids will tend to show more deviations since there are few steric constraints imposed upon them. Some structural aspects of a protein model can be verified using the methods based on fold recognition methods.

These methods rely on empirical pseudo-conormational energy potentials derived from the pairwise interactions observed in well-defined protein structures. These terms are summed over all residues in a model and result in a more (more negative) or less (more positive) favorable energy. These methods can detect a global sequence to structure incompatibility and errors corresponding to topological differences between template and the target. They also allow the detection of more localizes errors such as β-strands that are 'out of register' or buried charged residues. However, none of these methods has allowed the detection of more subtle structural inconsistencies, which are often localized in non-conserved loops, and can thus not provide an assessment of the correctness of their geometry.

Model Refinement and Evaluation Softwares

- There are a number of excellent software/tools for automatically evaluating protein structures and protein models. The important softwares are:
  1. DSSP (dictionary of secondary structure for proteins)
  2. PROCHECK
  3. WHATCHECK
  4. VADAR (The volume, area, dihedral angle, reporter)
  5. Verify 3D
  6. GROMOS (GROningen Molecular simulation package)
  7. ANOLEA (atomic non-local evaluation assessment)

Energy Minimization

Energy minimization (energy optimization) methods are common techniques to compute the equilibrium configuration of molecules. The basic idea is that a stable state of a molecular system should correspond to a local minimum of their potential energy. This kind of calculation generally starts from an arbitrary state of molecules, then the mathematical procedure of optimization allows us to move atoms (to vary variables) in a way to reduce the net forces (the gradients of potential energy) to nearly zero.

In computational chemistry, energy minimization (also called energy optimization or geometry optimization) methods are used to compute the equilibrium configuration of molecules and solids.

Stable state of molecular systems correspond to global and local minimum on their potential energy surface. Starting from a non-equilibrium molecular geometry, energy minimization employs the mathematical procedure of optimization to move atoms so to reduce the net forces (the gradients of potential energy) on the atoms until they become negligible.

Like molecular dynamics and Monte-Carlo approaches, periodic boundary conditions have been allowed in energy minimization methods, to make small systems. A well established algorithm of energy minimization can be an efficient tool for molecular structure optimization.

Unlike molecular dynamics simulations, which are based on Newtonian dynamic laws and allow calculating atomic trajectory with kinetic energy, molecular energy minimization does not include the effect of temperature, and hence the trajectories of atoms during the calculation do not really make any physical sense, i.e. we can only obtain a final state of system that corresponds to a local minimum of potential energy. From a physical point of view, this final state of the system corresponds to the configuration of atoms when the temperature of the system is approximately zero.

The algorithms of gradient are the most popular methods for energy minimization. The basic idea of gradient methods is to move atoms by the total net forces acting on them. The force on atoms is calculated as the negative gradient of total potential energy of system, as follows:

$$F(r_i) = -\overrightarrow{\nabla}_{ri} \cup^{tot}, \quad i = 1, \ldots, N,$$

where *ri* is the position of atom *i* and $\cup^{tot}$ is the total potential energy of the system.

10

An analytical formula of the gradient of potential energy is preferentially required by the gradient methods. If not, one needs to calculate numerically the derivatives of the energy function. In this case, the Powell's direction set method or the downhill simplex method can generally be more efficient than the gradient methods.

Simple Gradient Method

Here we have a single function of the potential energy to minimize with 3N independent variables, which are the 3 components of the coordinates of N atoms in our system. We calculate the net force on each atom F at each iteration step *t*, and we move the atoms in the direction of *F* with a multiple factor *k*. *k* can be smaller at the beginning of calculation if we begin with a very high potential energy. Note that similar strategy can be used in molecular dynamics for reducing the probability of divergence problems at the beginning of simulations.

$$r_i^t = r_i^{t-1} + \text{k.F}(r_i) \qquad \text{i} = 1, \ldots, \text{N}$$

We repeat this step in the above equation $t = 1, 2, \ldots$ until *F* reaches to zero for every atom. The potential energy of system goes down in a long narrow valley of energy in this procedure.

Despite that it is as well called "steepest descent", the simple gradient algorithm is in fact very time-consuming if we compare it to the nonlinear conjugate gradient approach, it is therefore known as a not very good algorithm. However, its advantage is its numerical stability, i.e. the potential energy can never increase if we take a reasonable *k*. Thus, it can be combined with a conjugated gradient algorithm for solving the numerical divergence problem when two atoms are too close to each other.

Nonlinear Conjugate Gradient Method

The conjugate gradient algorithm includes two basic steps—adding an orthogonal vector to the direction of research, and then move them in another direction nearly perpendicular to this vector. These two steps are as well known as—step on the valley floor and then jump down.

In this algorithm, we minimize the energy function by moving the atoms as follows,

$$r_i^t = r_i^{t-1} + \text{k.F}(r_i) \qquad \text{i} = 1, \ldots, \text{N}$$

where,

$$h_i^t = F(r_i^t) + \gamma_i^{t-1} h_i^{t-1}$$

and gamma is updated using the Fletcher-Reeves formula as:

$$\gamma_i^{t-1} = \frac{F(r_i^t).\ F(r_i^t)}{F(r_i^{t-1}).\ F(r_i^{t-1})}$$

Here we note that gamma can also be calculated by using the Polak-Ribiere formula, however, it is less efficient than the Fletcher-Reeves one for certain energy functions. At the beginning of calculation (when t = 1), we can make the search direction vector '$h0 = 0$.

This algorithm is very efficient. However, it is not quiet stable with certain potential functions, i.e. it sometimes can step so far into a very strong repulsive energy range (e.g. when two atoms are too close to each other), where the gradient on this point is almost infinite. It can directly result a typical data-overrun error during the calculation. For resolving this problem, we can combine the conjugated gradient algorithm with the simple one.

Applications of Energy Minimisation

Energy minimisation is very widely used in molecular modelling and is an integral part of techniques such as conformational search procedures. Energy minimisation is also used to prepare a system for other types of calculation. For example, energy minimisation may be used prior to a molecular dynamics or Monte Carlo simulation in order to relieve any unfavourable

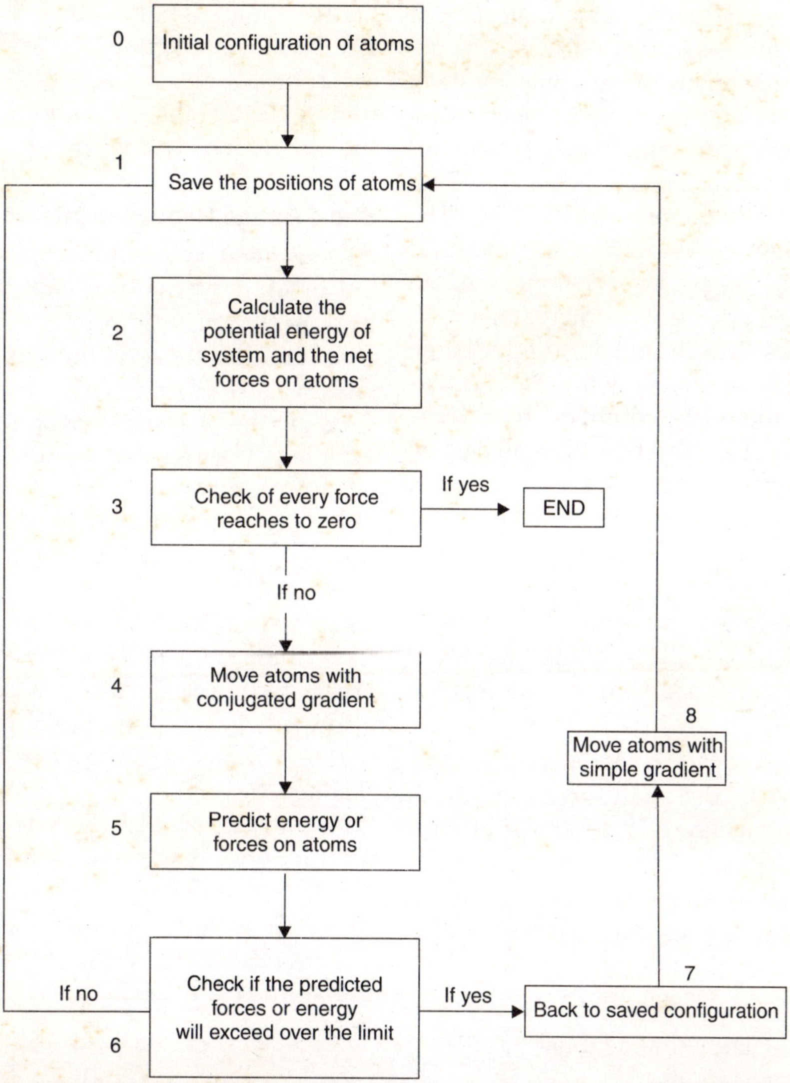

Fig. 10.1: Schematics of a computational energy minimization procedure

interactions in the initial configuration of the system. This is especially recommended for simulations of complex systems such as macromolecules or large molecular assemblies.

Molecular Interactions

Covalent interactions (bonds) hold the atoms of biopolymers and small molecules together. Covalent bond energies are on the order of 100 kcal/mole (400 joule/mole). Covalent bonds do not break when water boils, proteins fold (or unfold), when RNA folds, when DNA anneals, or when membranes assemble.

Those processes (protein folding/unfolding, etc.) are controlled by non-covalent interactions. Each given non-covalent interaction is relatively weak, often with a free energy of less than RT.

But cumulatively the energies are huge. Count the number of intramolecular pairwise interactions between the atoms of a globular protein, or between protein and water atoms in an unfolded protein. Huge numbers of small noncovalent forces drive the spontaneous folding or unfolding of proteins and nucleic acids. The folded/unfolded equilibrium of a proteins is generally small. The protein is held in 'delicate balance between powerful countervailing forces'. Large forces drive a protein to fold. Large forces drive it to unfold. It is the small difference between these large numbers that determines direction of the reaction. A small change in pH or temperature can change the balance. Unfortunately these forces are numerous, complex and poorly understood.

How to denature protein in your own kitchen? When you heat an egg to around 60 C, the proteins denature and aggregate (and that turns them white). Or add lemon juice to your milk.

Noncovalent forces were discovered by van der Waals during analysis of deviations from the ideal gas law. He noticed that molecules are sticky.

There are many different ways of classifying noncovalent interactions for our purposes. These are:

a. Repulsive interactions.
b. Charge—charge interactions.
c. Dipole interactions.
d. Fluctuating dipole.
e. Hydrogen bonding interactions.
f. Solvent, counter ion, and entropic effects.

Each of these is fundamentally electrostatic in nature and can described by Coulomb's law. Coulomb's law correctly describes forces that bind:

1. Electrons to nuclei in atoms,
2. Atoms to atoms in molecules, and
3. Molecules to molecules in liquids and solids.

One of our goals here is to avoid the term "van der Waals interaction" because that phrase incorporates all or many of those terms explicitly listed above, without decomposing the interactions in a meaningful way.

Short Range Repulsive Interactions

Atoms need space: Overlap of the occupied orbitals of two non-bonded atoms results in electrostatic repulsion between the electrons of those atoms. This repulsive energy acts over a very short range, but goes up very sharply when that range is violated. The repulsion goes up as $1/R^{12}$. It is important only when atoms are in very close proximity, but then it becomes very important. Because this repulsive term rises so sharply as distance decreases it is reasonable to think of atoms as hard spheres, like small pool balls, defined by van der Waals radii and surfaces. When two atoms approach each other their van der Waals surfaces make contact when their distance reaches the sum of their van der Waals radii (here we are assuming a lack of bonding interactions). The same principle applies to interactions between molecules although the shapes are more complex. The smallest distance between two non-bonded atoms is the sum of the van der Waals radii of the two atoms.

Charge–Charge Interactions

These are interactions between cations and anions, which are functional groups with formal charge. Charge–charge interactions are very strong, and cause the vapour pressure of sodium chloride, e.g. to be very low.

Dipole Interactions

In a molecule with unlike atoms, electrons are not shared equally. The tendency of any atom to pull electrons away from other atoms is characterized by a quantity called electronegativity.

Atom	Electronegativity
H	2.13
C	2.55
N	2.98
O	3.45
S	2.53

In a molecule composed of atoms of various electronegativities, the atoms with lowest electronegativities hold partial positive charges and the atoms with the greatest electronegativities hold partial negative charges.

Partial charges within a peptide bond.

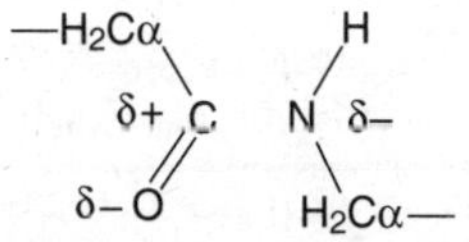

Fig. 10.2: Peptide bond

Atom	Partial charge (e)
N	−0.36
HN	+0.18
Ca	+0.06
C	+0.45
O	−0.38
Ha	+0.02

The extent of charge separation within a molecule is characterized by the dipole moment (m). To express dipole moments, charges are expressed in esu's and distances in centimeters. The dipole moment of an electron and a proton separated by 1 Å is given by:

$$(4.8 \;\; 10^{-10}\ \text{esu})\ (10^{-8}\,\text{cm})$$
$$= 4.8 \;\; 10^{-18}\ \text{esu cm} = 4.8\ \text{Debye}$$

The orientation of the dipole moment of a peptide is approximately parallel to the N–H bond and in magnitude is around 3.7 Debye. That is a relatively large dipole moment.

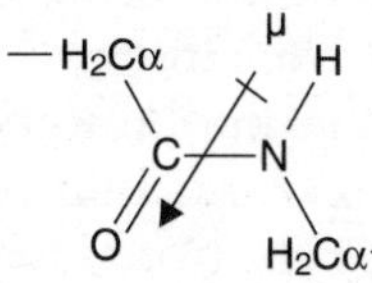

Fig. 10.3: Dipole moment of a peptide

The dipole moment of HCl is 1.0 Debye, that of CH_3Cl is 1.9 and that of HCN is 2.9. The large dipole moment of a peptide bond should lead one to expect that dipolar interactions are important in protein conformation and interactions. The large dipole of a peptide bond can be attributed in part to resonance.

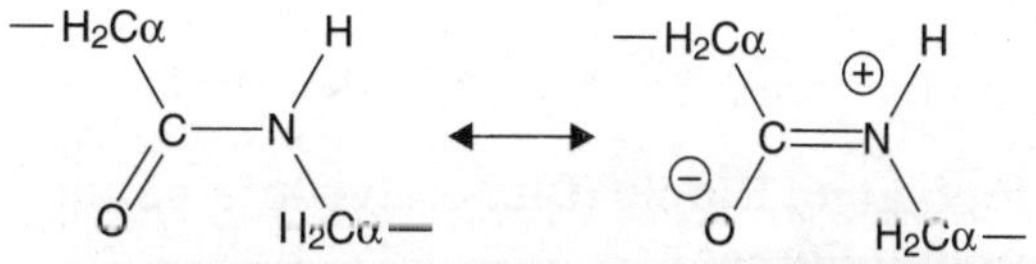

Fig. 10.4: Resonant structure of a peptide

A dipole can interact with point charges (called charge–dipole interaction), other dipoles (called dipole–dipole interaction), and can induce charge distribution in surrounding molecules (called dipole-induced dipole interaction). We will discuss each of these interactions separately.

Dipole–dipole interactions: The interaction energy between two dipoles can be either positive or negative and can be calculated with Coulomb's law. Listed below are the energies of interaction for two dipoles with moments of 1 Debye at a distance of 5 Å in a medium of $e = 4$.

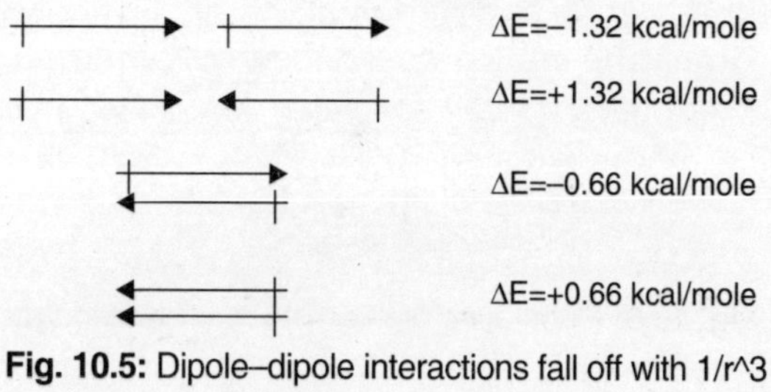

Fig. 10.5: Dipole–dipole interactions fall off with $1/r^3$

Dipole-induced dipoles interactions: A molecule with a permanent dipole can induce a dipole in a second molecule that is located nearby in space. The strength of the interaction depends on the dipole moment of the first molecule and the polarizability of the second. Molecules with pi electrons, such as benzene and phenyl alanine are more polarizable that molecules without pi electrons.

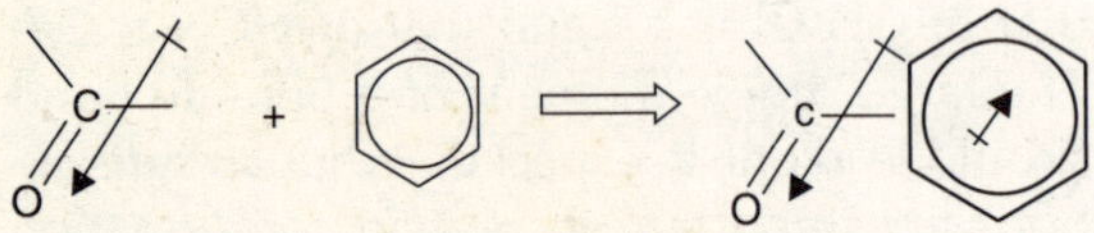

Dipole-induced dipole interactions are always attractive and can contribute as much as 0.5 kcal/mole to stabilization. Dipole-induced dipole interactions fall off with 1/r^4.

Fluctuating Dipole (Dispersive Interaction, London Forces)

Molecules behave like oscillating dipoles. In molecules that are located nearby to each other, the oscillators are coupled. Coupled fluctuating dipoles experience favorable electrostatic interaction known as dispersive interactions. The strength of the interaction is related to the polarizabilities of the two molecules.

Dispersive interactions are always attractive and occur between any pair of molecules, even non-polar molecules. Dispersive interactions provide the only attractive force between molecules in liquid N_2, which boils at 77K. For a given atom-atom contact, the energy of stabilization provided by dispersive interactions is very small (0.05 kcal/mole). However, the total number of contacts within a protein is generally enormous, so that dispersive interactions can make a large contribute to stability.

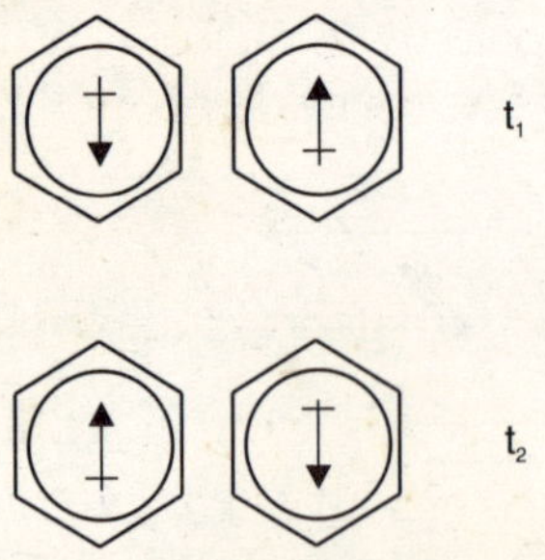

Fig. 10.6: Fluctuating dipoles of liquid benzene at two timepoints

Fluctuating dipole interactions fall off with 1/r^6.

Hydrogen Bonds

The idea that a single hydrogen atom could bond simultaneously to other two atoms was proposed in 1920 by Latimer and Rodebush and their advisor, G N Lewis. M L Huggins also describes the hydrogen bond in his 1919 dissertation.

An acceptor atom (A) with a basic lone pair of electrons can interact favorably with a donor atom (D) that bears an acidic proton.

A strong hydrogen bond requires that both atoms A and D are electronegative atoms. The most common hydrogen bonds in biological systems involve oxygen and nitrogen atoms. Sulfur can also engage in hydrogen bonds. Hydrogen bonds where atom D is a carbon atom have been observed although these are relatively weak interactions. Hydrogen bonds are essentially electrostatic in nature, although the energy can be decomposed into additional contributions from polarization, exchange repulsion, charge transfer, and mixing. In traversing the Period Table, increasing the electronegativity of atom D strips electron density from the proton, increasing its partial positive charge, and increasing the strength of the hydrogen bond. Two different ways of representing a hydrogen bond are shown below.

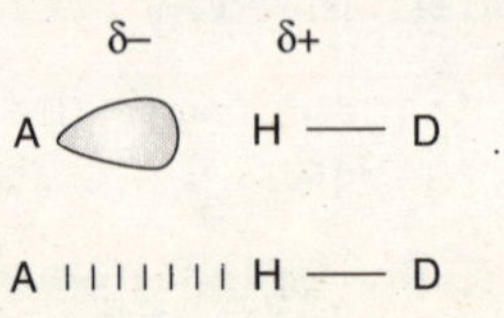

Fig. 10.7: Hydrogen bond representations

A hydrogen bond is not an acid-base reaction. In an acid-base reaction, but not in a hydrogen bond, the proton is transferred from D to A. But the acidity of the proton bound to D and the basicity of the lone pair of A both correlate roughly with the strength of the hydrogen bond.

Hydrogen bond strengths form a continuum. Strong hydrogen bonds of 20–40 kcal/mol, generally formed between charged donors and acceptors, are nearly as strong as covalent bonds. Weak hydrogen bonds of 1–5 kcal/mol, sometimes formed with carbon as the proton donor, are no stronger than van der Waals interactions. Moderate hydrogen bonds, which are the most common are formed between neutral donors and acceptors are from 5–15 kcal/mol.

The geometry of a hydrogen bond can be geometrically described by three quantities, the D to H distance, the H to A distance, and the D to H to A angle. Hydrogen bonds are not necessary, or even generally, linear.

Hydrogen atoms are not observable by X-ray crystallography as applied to proteins and nucleic acids. So a geometric description of hydrogen bonding that is dependent on proton position is not practical in protein and nucleic acid structures. In these cases, one is usually limited to analysis of the D to A distance.

Solvent, Counter Ion, and Entropic Effects

Nucleic acids in aqueous solutions can be considered rod-like polyanions surrounded by inorganic cations (Na^+, K^+ and Mg^{2+}) and polyamines. The high axial density of negatively charged phosphate groups on DNA causes strong radial electric fields. These electric fields lead to steep radial gradients of the surrounding cation concentrations. Theoretical considerations (counterion condensation) predict that the local concentration of a monovalent cation such as K^+ near the surface of DNA is around 2 molar. This local concentration (i.e. near the DNA) of K^+ is largely independent of K^+ concentration in bulk solution.

Release or uptake of local cations accompanies processes such as helix-coil and other conformational transitions that change the axial charge density. Release or uptake of local cations also accompanies binding of proteins, and intramolecular compaction or collapse of DNA. These effects explain the dramatic salt dependencies of DNA melting and DNA-protein complexation.

DNA condensation: Genomic DNAs are very long molecules. The 160,000 base pairs of T_4 phage DNA extend to 54 microns. The 4.2 million base pairs of the *E. coli* chromosome extend to 1.4 millimeters. In biological systems, long DNA molecules must be compacted to fit into very small spaces inside a cell, nucleus or virus particle. The energetic barriers for tight packaging of DNA arise from decreased configurational entropy, bending the stiff double helix, and intermolecular (or intersegment) electrostatic repulsion of the negatively charged DNA phosphate groups. Yet extended DNA chains condense spontaneously by collapsing into very compact, orderly particles. In the condensed state, DNA helixes are separated by one or two layers of water. Condensed DNA particles are commonly compact toroids. DNA condensation in aqueous solution requires highly charged cations such as spermine (+4) or spermidine (+3). Divalent cations will condense DNA in water-alcohol mixtures. The role of the cations is to decrease electrostatic repulsion of adjacent negatively charged DNA segments. The source of the attraction between nearby DNA segments is not so easy to understand. One possible source of attraction are fluctuations of ion atmospheres in analogy with fluctuating dipoles between molecules (London Forces).

Proteins: When thinking about the free energy of protein folding (folding is the transition from denatured state to native state) it is important to understand that the stability of a folded state in a biological system can only be understood as a difference between the free energies of the native and denatured states.

DGfolding = DGnative state–DGdenatured state

The molecular interactions described in the previous sections (charge–charge, dipole-dipole, hydrogen bonding...) are observed in both states. In the native state, many of these interactions are intramolecular, e.g. one part of the protein will form a hydrogen bond with another part of the protein. In the denatured state, interactions are intermolecular, (i.e. between the protein and water molecules, cations, and anions). A intramolecular hydrogen bond observed in the native state will be replaced in the denatured state by several hydrogen bonds between the protein and water molecules. The same is essentially true for charge–charge, dipole–dipole, dipole-induced, and fluctuations dipole interactions.

One of the primary driving forces for protein folding is solvent release. In the transition from denatured to native state, a large number of solvent molecules are released. Release is defined as the conversion from bound state to free state.

Force Field

In the context of molecular mechanics, a force field (also called a forcefield) refers to the functional form and parameter sets used to describe the potential energy of a system of particles (typically but not necessarily atoms). Force field functions and parameter sets are derived from both experimental work and high-level quantum mechanical calculations. "All-atom" force fields provide parameters for every atom in a system, including hydrogen, while "united-atom" force fields treat the hydrogen and carbon atoms in methyl and methylene groups as a single interaction center. "Coarse-grained" force fields, which are frequently used in long-time simulations of proteins, provide even more abstracted representations for increased computational efficiency.

The usage of the term "force field" in chemistry and computational biology differs from the standard usage in physics. In chemistry usage, a force field is defined as a potential function, while the term is used in physics to denote the negative gradient of a scalar potential.

Functional Form

The basic functional form of a force field encapsulates both bonded terms relating to atoms that are linked by covalent bonds, and nonbonded (also called "noncovalent") terms describing the long-range electrostatic and van der Waals forces. The specific decomposition of the terms depends on the force field, but a general form for the total energy in an additive force field can be written as:

$$E_{total} = E_{bond} + E_{nonbonded}$$

where the components of the covalent and noncovalent contributions are given by the following summations:

$$E_{bonded} = E_{bond} + E_{dihedral}$$

$$E_{nonbonded} = E_{electrostatic} + E_{vanderWalls}$$

The bond and angle terms are usually modelled as harmonic oscillators in force fields that do not allow bond breaking. A more realistic description of a covalent bond at higher stretching is provided by the more expensive Morse potential. The functional form for the rest of the bonded terms is highly variable. Proper dihedral potentials are usually included. Additionally, "improper torsional" terms may be added to enforce the planarity of aromatic rings and other conjugated systems, and "cross-terms" that describe coupling of different internal variables, such as angles and bond lengths. Some force fields also include explicit terms for hydrogen bonds.

The nonbonded terms are most computationally intensive because they include many more interactions per atom. A popular choice is to limit interactions to pairwise energies. The van der Waals term is usually computed with a Lennard-Jones potential and the electrostatic term with Coulomb's law, although both can be buffered or scaled by a constant factor to account for electronic polarizability and produce better agreement with experimental observations.

Parameterization

In addition to the functional form of the potentials, a force field defines a set of parameters for each type of atom. For example, a force field would include distinct parameters for an oxygen atom in a carbonyl functional group and in a hydroxyl group. The typical parameter set includes values for atomic mass, van der Waals radius, and partial charge for individual atoms, and equilibrium values of bond lengths, bond angles, and dihedral angles for pairs, triplets, and quandruplets of bonded atoms, and values corresponding to the effective spring constant for each potential. Most current force fields use a "fixed-charge" model by which each atom is assigned a single value for the atomic charge that is not affected by the local electrostatic environment; proposed developments in next-generation force fields incorporate models for polarizability, in which a particle's charge is influenced by electrostatic interactions with its neighbors. For example, polarizability can be approximated by the introduction of induced dipoles; it can also be represented by Drude particles, or massless, charge-carrying virtual sites attached by a spring-like harmonic potential to each polarizable atom. The introduction of polarizability into force fields in common use has been inhibited by the high computational expense associated with calculating the local electrostatic field.

Although many molecular simulations involve biological macromolecules such as proteins, DNA, and RNA, the parameters for given atom types are generally derived from observations on small organic molecules that are more tractable for experimental studies and quantum calculations. Different force fields can be derived from dissimilar types of experimental data, such as enthalpy of vaporization (OPLS), enthalpy of sublimation (CFF), dipole moments, or various spectroscopic parameters (CFF).

Parameter sets and functional forms are defined by force field developers to be self-consistent. Because the functional forms of the potential terms vary extensively between even closely related force fields (or successive versions of the same force field), the parameters from one force field should never be used in conjunction with the potential from another.

Classical force fields

- AMBER (assisted model building and energy refinement)] — widely used for proteins and DNA.
- CHARMM (chemistry at HARvard molecular mechanics)] — originally developed at Harvard, widely used for both small molecules and macromolecules.
- CHARMm — commercial version of CHARMM, available through Accelrys
- CVFF - also broadly used for small molecules and macromolecules.
- COSMOS-NMR—hybrid QM/MM forcefield adapted to a variety of inorganic compounds, organic compounds and biological macromolecules, including semiempirical calculation of atomic charges and NMR properties. COSMOS-NMR, is optimized for NMR-based structure elu-

cidation and implemented in COSMOS molecular modelling package.

- GROMACS — the force field optimized for the package of the same name.
- GROMOS—a force field that comes as part of the GROMOS (GROningen MOlecular simulation package), a general-purpose molecular dynamics computer simulation package for the study of biomolecular systems. GROMOS force field (A-version) has been developed for application to aqueous or apolar solutions of proteins, nucleotides and sugars. However, a gas phase version (B-version) for simulation of isolated molecules is also available.
- OPLS (optimized potential for liquid simulations) (variations include OPLS-AA, OPLS-UA, OPLS-2001, OPLS-2005) developed by William L. Jorgensen at the Yale University Department of Chemistry.
- ENZYMIX — a general polarizable force field for modelling chemical reactions in biological molecules. This force field is implemented with the empirical valence bond (EVB) method and is also combined with the semimacroscopic PDLD approach in the program in the MOLARIS package.
- ECEPP/2 — first force field for polypeptide molecules — developed by FA Momany, HA Scheraga and colleagues.
- QCFF/PI — a general force field for conjugated molecules.

Second-generation force fields

- CFF—a family of forcefields adapted to a broad variety of organic compounds, includes force fields for polymers, metals, etc.
- MMFF (Merck molecular force field)-developed at Merck, for a broad range of molecules.
- MM2, MM3, MM4 — developed by Norman Allinger, parametrized for a broad range of molecules.
- QVBMM — developed by Vernon GS Box, parameterized for all biomolecules and a broad range of organic molecules, and implemented in STR3DI32.

Polarizable force field based on induced dipole

- CFF/ind and ENZYMIX – the first polarizable force field which has subsequently been used in many applications to biological systems.
- DRF90 developed by P. Th. van Duijnen and coworkers.

Polarizable force fields based on point charges

- PFF (polarizable force field) developed by Richard A. Friesner and coworkers.
- SP-basis chemical potential equalization (CPE) approach developed by R. Chelli and P. Procacci.
- CHARMM polarizable force field developed by S. Patel (University of Delaware) and C L Brooks III (University of Michigan).
- AMBER polarizable force field developed by Jim Caldwell and coworkers.

Polarizable force fields based on distributed multipoles

- The SIBFA (sum of interactions between fragments Ab initio computed) force field for small molecules and flexible proteins, developed by Nohad Gresh (Paris V, René Descartes University) and Jean-Philip Piquemal (Paris VI, Pierre & Marie Curie University). SIBFA is a molecular mechanics procedure formulated and calibrated on the basis of ab initio supermolecule computations. Its purpose is to enable the simultaneous and reliable computations of both intermolecular and conformational energies governing the binding specificities of biologically and pharmacologically relevant molecules. This procedure enables an accurate treatment of transition metals. The inclusion of a ligand field contribution allows computations on "open-shell" metalloproteins.

- AMOEBA (atomic multipole optimized energetics for biomolecular applications) force field developed by Pengyu Ren (University of Texas at Austin) and Jay W. Ponder (Washington University).
- ORIENT procedure developed by Anthony J. Stone (Cambridge University) and coworkers.
- Non-empirical molecular orbital (NEMO) procedure developed by Gunnar Karlström and coworkers at Lund University (Sweden).

Polarizable force fields based on bond polarization theory (BPT)

- COSMOS-NMR (computer simulation of molecular structure) developed by Ulrich Sternberg and coworkers. Hybrid QM/MM force field enables explicit quantum-mechanical calculation of electrostatic properties using localized bond orbitals with fast BPT formalism. Atomic charge fluctuation is possible in each molecular dynamics step.

Reactive force fields

- ReaxFF-reactive force field developed by Adri van Duin, William Goddard and coworkers. It is fast, transferable and is the computational method of choice for atomistic-scale dynamical simulations of chemical reactions.
- EVB (empirical valence bond) – this reactive force field, introduced by Warshel and coworkers, is probably the most reliable and physically consistent way of using force fields in Modelling chemical reactions in different environments. The EVB facilitates calculations of actual activation free energies in condensed phases and in enzymes.
- RWFF - reactive force field for water developed by Detlef WM Hofmann, Liudmila N Kuleshova and Bruno D'Aguanno. It is very fast, reproduces the experimental data of neutron scattering accurate, and allows to simulate the bond formation/breaking of water and acids.

Coarse-grained force fields

- VAMM (virtual atom molecular mechanics) force field developed by Korkut and Hendrickson. VAMM is a coarse-grained force field specifically designed for molecular mechanics calculations such as large scale conformational transitions based on the virtual interactions of Calpha atoms. It is a knowledge-based force field and formulated to capture features dependent on secondary structure and on residue-specific contact information in proteins.

Other

- VALBOND — a function for angle bending that is based on valence bond theory and works for large angular distortions, hypervalent molecules, and transition metal complexes. It can be incorporated into other force fields such as CHARMM and UFF.

Molecular Dynamics

Molecular dynamics (MD) is a form of computer simulation in which atoms and molecules are allowed to interact for a period of time by approximations of known physics, giving a view of the motion of the atoms. Because molecular systems generally consist of a vast number of particles, it is impossible to find the properties of such complex systems analytically; MD simulation circumvents this problem by using numerical methods. It represents an interface between laboratory experiments and theory, and can be understood as a "virtual experiment". MD solves the relationship between molecular structure, movement and function. Molecular dynamics is a multidisciplinary method. Its laws and theories stem from mathematics, physics, and chemistry, and it employs algorithms from computer science and information theory. It was originally conceived within theoretical physics in the late 1950s, but is applied today mostly in materials science and biomolecules.

Molecular dynamics calculates the 'real' dynamics of the system, from which time averages of properties can be calculated. Sets of atomic positions are derived in sequence by applying Newton's equations of motion. Molecular dynamics is a deterministic method, by which we mean that the state of the system at any future time can be predicted from its current state. The first molecular dynamics simulations were performed using very simple potentials such as the hard-sphere potential. The behaviour of the particles in this potential is similar to that of billiard or snooker balls, the particles move in straight lines at constant velocity between collisions. The collisions are perfectly elastic and occur when the separation between a pair of spheres equals the sum of their radii. After a collision, the new velocities of the colliding spheres are calculated using the principle of conservation of linear momentum. The hard sphere model has provided many useful results but is obviously not ideal for simulating atomic or molecular systems. In potentials such as the Lennard-Jones potential the force between two atoms or molecules changes continuously with their separation. By contrast, in the hard-sphere model there is no force between particles until they collide. The continuous nature of the more realistic potentials requires the equations of motion to be integrated by breaking the calculation into a series of very short time steps (typically between 1 femto-femtosecond and 10 femtoseconds; 10–15 s to 10–14 s). At each step, the forces on the atoms are computed and combined with the current positions and velocities to generate new positions and velocities a short time ahead. The force acting on each atom is assumed to be constant during the time interval. The atoms are then moved to the new positions, an updated set of forces is computed, and so on. In this way, a molecular dynamics simulation generates a trajectory that describes how the dynamic variables change with time.

Application of MD algorithms

In chemistry, MD serves as an important tool in protein structure determination and refinement using experimental tools such as X-ray crystallography and NMR. It has also been applied with limited success as a method of refining protein structure predictions. In physics, MD is used to examine the dynamics of atomic-level phenomena that cannot be observed directly, such as thin film growth and ion-subplantation. It is also used to examine the physical properties of nanotechnological devices that have not or cannot yet be created.

In applied mathematics and theoretical physics, molecular dynamics is a part of the research of dynamical systems, ergodic theory and statistical mechanics in general. The concepts of energy conservation and molecular entropy come from thermodynamics. Some techniques to calculate conformational entropy such as principal components analysis come from information theory. Mathematical techniques such as the transfer operator become applicable when MD is seen as a Markov chain. Also, there is a large community of mathematicians working on volume preserving, symplectic integrators for more computationally efficient MD simulations.

MD can also be seen as a special case of the discrete element method (DEM) in which the particles have spherical shape (e.g. with the size of their van der Waals radii).

Molecular Modelling in Drug Discovery

Most drugs produce their effect by interacting with a biological macromolecule such as an enzyme, DNA, glycoprotein or receptor. The interaction between a ligand and its target may be due entirely to non-bonded forces, but in some cases a covalent interaction may be involved. Drugs which interact with receptor proteins can be classified as agonists, antagonists or inverse agonists. Agonists produce the same or elevated effect as the natural substrate or effector molecule, whereas antagonists inhibit the effect of the natural ligand. Inverse agonists create an effect which appears opposite to that of the agonist. Tight-binding ligands often have a high degree of complementarity with the target. This complementarity can be assessed and measured in various ways. Many ligands show significant shape complementarity with the region of the macromolecule where they bind (the binding site). The ligand often forms hydrogen bonds with the receptor. Some receptors have hydrophobic 'pockets', formed by groups of non-polar amino acids, into which the ligand can place a hydrophobic group of an appropriate size. It is also crucial to remember that a good drug does more than simply bind tightly to its target. After administration, a drug must get to the site of action. This transport process often requires the drug to pass through cell membranes. A cell membrane is a hydrophobic environment and so the drug must be sufficiently lipophilic (lipid loving) to partition into the membrane but not so lipophilic that it stays there. Once inside the cell, the drug must access its target. During this process, the molecule may also be removed from the body by metabolism, excretion and other pathways.

Discovering and developing any new medicine is a long and expensive process. A new compound must not only produce the desired response with minimal side effects but must also be demonstrably better than existing therapies. Two key steps in many drug discovery programmes are the identification of hit molecules ('hits') and lead series ('leads'). A hit is a molecule that has some reproducible activity in a biological assay. A lead series comprises a set of related molecules that usually share some common structural feature, and which show some variation in the activity as the structure is modified. This gives one confidence that further synthetic modification to the lead sites (termed lead optimisation) has a good chance of resulting in a drug candidate with the desired potency and selectivity, lack of toxicity and the appropriate characteristics to enable it to reach its target in vivo. Such a drug candidate will then enter the early stages of development, where further large-scale investigations are undertaken.

Finding novel lead series can be a difficult problem. Serendipity often played an important role in the past, a classic example being the discovery of penicillin by Alexander Fleming. For many years, pharmaceutical companies have screened soil and other biological samples to find new leads, but it can be difficult to extract and purify the bioactive ingredient. The 1990s saw the widespread adoption of high-throughput screening (HTS), which enables large numbers of compounds to be screened using highly automated, robotic techniques. The molecules used for HTS come from compounds synthesised in earlier medicinal chemistry programmes, from compounds that can be purchased from chemical suppliers and from combinatorial chemistry. However, although HTS makes it possible in principle to test every available compound against every biological assay, there are a number of practical reasons why this is not necessarily feasible, let alone desirable. The first reason is financial—although robotics and miniaturisation have significantly reduced the unit cost, the sheer number of samples now available in many companies means that the overall expense can be significant. A second reason is that some assays cannot be converted to a high-throughput format and so have to be conducted using more traditional techniques. Third, a significant proportion of the available samples might not be considered appropriate structures, suitable for taking forward to the next stage. For example, some molecules may contain functional groups which are known to react in a non-specific manner with biological targets. Other molecules might interfere with the proper interpretation of the assay, such as a strong fluorophore. Yet more molecules may just be considered 'inappropriate', or not sufficiently 'drug-like'.

For these and other reasons, it is often necessary to identify subsets of compounds. Computational techniques have a significant role to play in the ways in which such subsets can be constructed, with various techniques being available depending upon the type of molecule that one wishes to screen, what kind of information is available to assist the selection and what properties one wishes to take into account.

Introduction to Molecular Docking

Molecular docking is a method to predict the preferred orientation of one molecule to a second when bound to each other to form a stable complex. Computers and programs (softwares) are used to predict or simulate the possible reaction (and interactions) between two molecules based on their three-dimensional structures. Using softwares, the interactions can be viewed and analyzed to understand and answers some biological important questions regarding a certain chemical or biological reaction. Analyzing the interactions basically comes with (nice) 3D graphics which can be manipulated in several ways to clearly explore in detail (in atomic resolutions) the interaction involved between the atoms in the two interacting molecules. This method can therefore be used not only to predict possible binders or inhibitors, but also to predict how strong the association between the molecules (called the binding affinity) can be. It is useful to know the binding strength (binding energy) when you are comparing (ranking) a group of compounds or derivatives to determine which derivative is the best binder or inhibitor (how strong a compound will bind to the target).

Prediction of the binding affinity will be useful when you are synthesizing compounds whereby you can predict the affinity of your desired compound towards a certain target (say a protein or DNA; with particular interest to stop the function of the enzyme/protein or to block certain reaction).

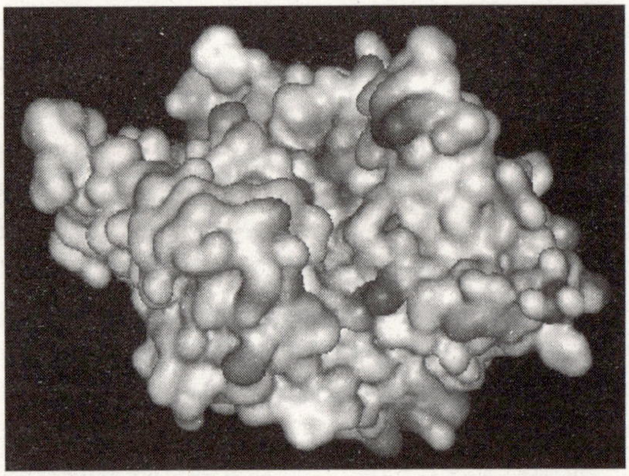

Fig. 10.8: Molecular surface of a protein

You can therefore save a lot time and money by "experimenting" using the computer first before actually going to the lab to make your compound. In addition, you can predict how a molecule interact or react with another moleculem, e.g. in protein–protein interaction, in a specific biological reaction (of your interest) before conducting the ("wet") experiment.

This method is also useful when you want to screen (they call it "virtual screening") a number of compounds say from a natural product or plants/herbs to see whether your small molecules (from the medicinal plants/herbs) will have certain pharmacological effects on a particular protein or enzyme, e.g. HIV protease etc.). Large pharmaceutical companies in Europe and US have been using this technique for some time in the discovery and development of new drugs.

There are two main types of docking (molecular docking) in practice: small molecule–protein (called "ligand–protein docking") and protein–protein docking. As mentioned earlier, there is also small molecule – DNA or RNA docking done by some researchers. I believe, this can be categorized as ligand–DNA/RNA docking.

Protein–protein docking involves two protein molecules simulated by the computer/computer program to bind/interact with one another. However, in this case, the interactions are basically rigid compared to the ligand–protein docking. You might be able to see that by simulating certain protein–protein interactions in a specific biological reaction, you can get some information or insights at the molecular level on how a certain mechanism took place.

There are many docking programs available which are able to simulate the ligand protein interaction with the ligand (small molecule) having full flexibility (rotations in their bonds) while the protein or receptor either rigid or partially flexible (some amino acid side chains on the interaction site of the protein were made to be able to rotate). The higher the flexibility, the more computational time needed to simulate the interaction.

Why is it important?

Because of its ability in predicting binding interactions and orientation (in some cases at a very high accuracy with reference to existing crystal structure of the complex studied), it is being widely used in rational drug design and structure-based drug design processes (structure-based drug design means that we use three-dimensional structures to design drug/new drugs with the help of computers and softwares).

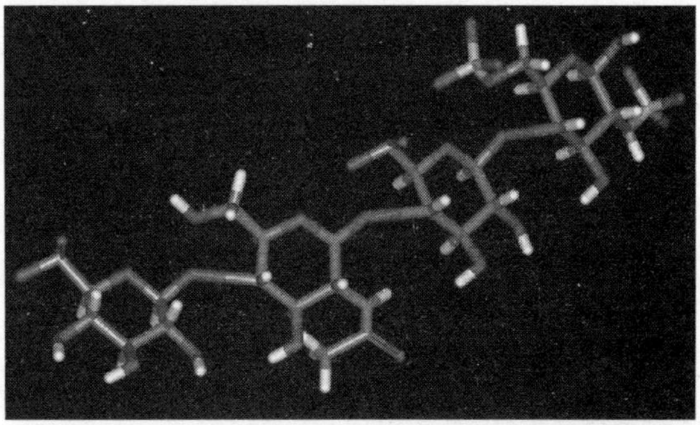

Fig. 10.9: 3D structure of a ligand with many torsions

Another good reason why many researchers are moving towards docking methods (to complement their work) in their research is because some information are difficult to obtain through experimental ways. The ability of the computer to simulate the reactions in atomic details (nanoscale) and with the increasing power of computers (high performance computing), provides the answer to difficult research problems which cannot be solved through the conventional means. This is why many researchers have diverted some (if not all) of their attention into this technique.

Brief Theory

The theory behind molecular docking lies behind the enzyme–substrate recognition process. The problem can be thought as a "lock – and – key" concept. In this problem, the orientation of the ligand (small molecule or substrate protein) will be "fitted" to the receptor of interest using either 2 approaches; matching technique, and simulation processes. Certain programs (softwares) are able to rank the affinity of compounds towards the receptor studied. The programs will "adjust" the conformation of the ligand and in some cases the conformation of the side chains of the receptors binding site (site of interaction) to see how they will fit each other. This is done by calculating the energy for each conformation of the complex and the molecules when they are in their uncomplex form (before binding).

Methods of Calculation

1st method

The first methods uses matching technique. This technique looks for complementary of the surfaces of each of the molecules (solvent–accessible surface area) to see whether it will match each other surface. This is called the shape complementary methods.

2nd method

The second method uses simulation processes. This technique simulates the actual docking process by calculating the ligand–protein pairwise interaction energies (using emperical methods or molecular mechanics—take into account torsional energy between bonds, van der Waals energy, electrostatic energy, hydrogen bonding potential, atomic solvation energy, and so forth).

How Docking is Done?

Typically, you will need two types of file or input; one for the ligand/small molecule and another one for the target/receptor. The ligand or small molecule can be built from scratch using suitable software or taken from available databases.

Some software comes with a two-dimensional drawing which can be converted to three-dimensional type. There are softwares with three-dimensional drawing capabilities. The protein or receptor's structure can either be downloaded from several databases, for example protein databank (http://www.rcsb.org/) or you can build it based on a template using softwares with amino acids/biopolymer construction features (called homology modelling). The protein databank is a good resiporatory of large collections of X-ray crystallographic and NMR (nuclear magnetic resonance) 3D structures available for the public.

You will need to check the structure built (the receptor) to validate the quality of your structure before performing any docking. You can easily find programs to do this over the web. So now you have 2 inputs (in 3D) needed to perform your docking/simulation. One for the ligand, and the other one for your target/receptor.

How reliable is a docking result?

There are many docking programs available on the net, commercial or open source. However, the success of a program mainly depends on two components: the search algorithm and, the scoring function. Of course, its only a computer, if you put "rubbish in", you get "rubbish out". You will need to carefully check all your inputs and parameters before accepting any results from the simulation. The resolution of the receptor used also play an important part in the accuracy of your simulation. The higher the resolution of your protein or receptor, the higher the accuracy of your result.

The Search Algorithm

The search algorithm is a process where all possible conformations and orientations of the complex (the paired ligand and protein) in a space (the binding site of interest) is being searched. If the ligand is flexible, then the program will calculate the energy for each rotation made of each and every rotatable bonds it can find. The same goes for the protein/receptor. For each and every rotation of the side chain of the amino acids (in the binding site), the program will calculate the energy involved. Each energy value calculated will be presented as a "snapshot" of the pair.

In each of the snapshot, you will be able to see what kind of interactions are involved and which atoms are in contact or close proximity making any kinds of bonding such as hydrogen bonds, hydrophobic interactions and many more. Each of the "snapshot" of the pair or the complex is called the binding "pose".

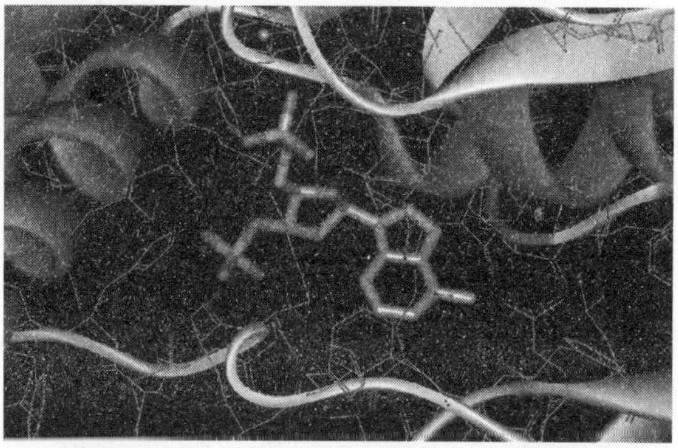

Fig. 10.10: Example: A binding pose of the docked ligand

The Scoring Function

The scoring function is a process where the program takes a binding pose and gives a number to indicate the likelihood whether the binding interaction is favorable. Molecular mechanics force field (physics-based calculations) is used to estimate the energy of each pose. Each and every pose will come with an energy value and the scoring function of the program will rank the poses accordingly (normally in a descending manner). The lower the energy of the pose, the more stable the complex will be, and more likely the possibility of the binding will happen.

Application of Docking

As mentioned earlier, most application goes to the field of drug design. Most drugs, as we know are small organic molecules. Another application is in the prediction of the activation or inhibition of a particular enzyme or protein (or in some cases DNA) of interest.

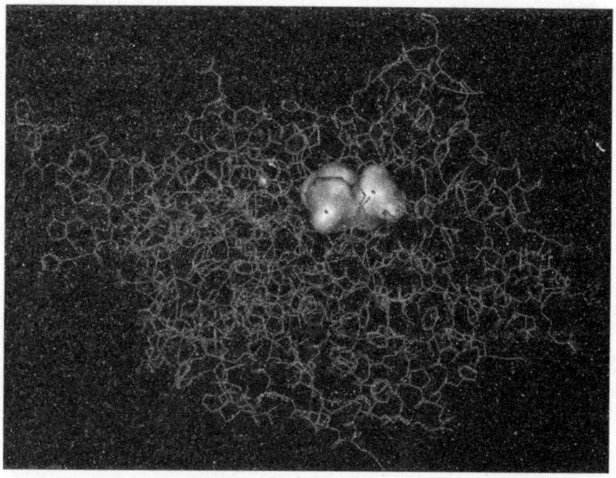

Fig. 10.11: A ligand in the cavity/active site of a protein

In the process of identifying potential drug candidates from large databases of drugs/ small molecules, many efforts have been made to identify molecules with tendency to bind to a protein target of interest. This process is sometimes called virtual screening or hits identification or insilico drug screening where thousands of drug candidates are being screened rapidly using high speed or high performance computing facilities.

Docking can also be used in "lead optimization" process. This is a process where the lead or drug candidate which had earlier shown binding affinity towards the protein, is being structurally modified (in the computer) to enhance its binding potential. The potency and/or the selectivity of the drug candidate towards the protein, can therefore be improved. Researchers have been using this technique lately as part of an effort to help minimize the problems with late failure in drug development. 10

Docking software

There are several computational packages that are available for docking. Generally, all of them involve GRID generation and scoring techniques. The result of docking is usually inferred in terms of energy. The docked complex with lowest energy is considered as a stable complex. The important computational tools for docking as follows:

- DOCK
- AutoDOCK
- FTDOCK
- ZDOCK
- 3D DOCK
- Argus lab
- HEX
- CHARMM
- Affinity

Applications of Molecular Docking

A binding interaction between a small molecule ligand and a enzyme protein may result in activation or inhibition of the enzyme. If the protein is a receptor, ligand binding may result in agonism or antagonism. Docking is most commonly used in the field of drug design — most drugs are small organic molecules, and docking may be applied to:

1. *Hit identification:* Docking combined with a scoring function can be used to quickly screen large databases of potential drugs in silico to identify molecules that are likely to bind to protein target of interest (virtual screening).
2. *Lead optimization:* Docking can be used to predict in where and in which relative orientation a ligand binds to a protein (also referred to as the binding mode or pose). This information may in turn be used to design more potent and selective analogs.
3. *Bioremediation:* Protein ligand docking can also be used to predict the putative pollutants that can be degradable by the biodegradative proteins.
4. Docking is used for structure-based drug discovery, where the binding efficiency of particular drug is computationally tested by docking it with the known well studied target structure.
5. Docking could also used to study protein–protein interaction.

Quantitative Structure-activity Relationship (QSAR)

Quantitative structure-activity relationship (QSAR) (sometimes QSPR: quantitative structure -property relationship) is the process by which chemical structure is quantitatively correlated with a well-defined process, such as biological activity or chemical reactivity.

For example, biological activity can be expressed quantitatively as in the concentration of a substance required to give a certain biological response. Additionally, when physicochemical properties or structures are expressed by numbers, one can form a mathematical relationship, or quantitative structure-activity relationship, between the two. The mathematical expression can then be used to predict the biological response of other chemical structures.

QSARs most general mathematical form is:

- Activity = *f* (physiochemical properties and/or structural properties)

SAR and the SAR Paradox

The basic assumption for all molecule-based hypotheses is that similar molecules have similar activities. This principle is also called structure-activity relationship (SAR). The underlying problem is therefore how to define a small difference on a molecular level, since each kind of activity, e.g. reaction ability, biotransformation ability, solubility, target activity, and so on, might depend on another difference.

In general, one is more interested in finding strong trends. Created hypotheses usually rely on a finite number of chemical data. Thus, the induction principle should be respected to avoid overfitted hypotheses and deriving overfitted and useless interpretations on structural/molecular data.

The SAR paradox refers to the fact that it is not the case that all similar molecules have similar activities.

Types

Fragment-based (group contribution): It has been shown that the logP of compound can be determined by the sum of its fragments. Fragmentary logP values have been determined statistically. This method gives mixed results and is generally not trusted to have accuracy of more than ±0.1 units.

3D-QSAR: 3D-QSAR refers to the application of force field calculations requiring three-dimensional structures, e.g. based on protein crystallography or molecule superposition. It uses computed potentials, e.g. the Lennard-Jones potential, rather than experimental constants and is concerned with the overall molecule rather than a single substituent. It examines the steric fields (shape of the molecule) and the electrostatic fields based on the applied energy function.

In the literature, it can be often found that chemists have a preference for partial least squares (PLS) methods, since it applies the feature extraction and induction in one step.

Data mining

For the coding, usually a relatively large number of features or molecular descriptors is calculated, which can lack structural interpretation ability. In combination with the later applied learning method or as preprocessing step occurs a feature selection problem.

A typical data mining-based prediction uses, e.g. support vector machines, decision trees, neural networks for inducing a predictive learning model.

Molecule mining approaches, a special case of structured data mining approaches, apply a similarity matrix-based prediction or an automatic fragmentation scheme into molecular substructures. Furthermore there exist also approaches using maximum common subgraph searches or graph kernels.

Planning a QSAR study

When starting a QSAR study, it is important to decide which physicochemical parameters are going to be studied and to plan the analogues such that the parameters under study are suitably varied. For example, it would be pointless to synthesize analogues where the hydrophobicity and steric volume of the substituents are correlated, if these two parameters are to go into the equation.

It is also important to make enough structures to make the results statistically meaningful. As a rule of thumb, five structures should be made for every parameter studied. Typically, the initial QSAR study would involve the two parameters and σ, and possibly *Es*.

Certain substituents are worth avoiding in the initial study since they may have properties other than those being studied. For example, substituents which might ionize (CO_2H, NH_2, SO_2H) should be avoided. Groups which might easily be metabolized should be avoided if possible (e.g. esters or nitro groups).

If there are two or more substituents, then the initial equation usually considers the total and σ contribution. As more analogues are made, it is often possible to consider the hydrophobic and electronic effect of substituents at specific positions of the molecule. Furthermore, the electronic parameter a can be split into its inductive and resonance components (F and R). Such detailed equations may show up a particular localized requirement for activity. For example, a hydrophobia substituent may be favoured in one part of the skeleton, while an electron withdrawing substituent is favoured at another. This in turn gives clues about the binding interactions involved between drug and receptor.

Judging the quality of QSAR models

QSARs represent predictive models derived from application of statistical tools correlating biological activity (including desirable therapeutic effect and undesirable side effects) of chemicals (drugs/toxicants/environmental pollutants) with descriptors representative of molecular structure and/or properties. QSARs are being applied in many disciplines for example risk assessment, toxicity prediction, and regulatory decisions in addition to drug discovery and lead optimization. Obtaining a good quality QSAR model depends on many factors, such as the quality of biological data, the choice of descriptors and statistical methods. Any QSAR modelling should ultimately lead to statistically robust models capable of making accurate and reliable predictions of biological activities of new compounds.

For validation of QSAR models, usually four strategies are adopted:

1. Internal validation or cross-validation;
2. Validation by dividing the data set into training and test compounds;
3. True external validation by application of model on external data and
4. Data randomization or Y-scrambling.

The success of any QSAR model depends on accuracy of the input data, selection of appropriate descriptors and statistical tools, and most importantly validation of the developed model. Validation is the process by which the reliability and relevance of a procedure are established for a specific purpose. Leave one-out cross-validation generally leads to an overestimation of predictive capacity, and even with external validation, no one can be sure whether the selection of training and test sets was manipulated to maximize the predictive capacity of the model being published. Different aspects of validation of QSAR models that need attention includes methods of selection of training set compounds, setting training set size and impact of variable selection for training set models for determining the quality of prediction. Development of novel validation parameters for judging quality of QSAR models is also important.

Application

Biological: The biological activity of molecules is usually measured in assays to establish the level of inhibition of particular signal transduction or metabolic pathways. Chemicals can also be biologically active by being toxic. Drug discovery often involves the use of QSAR to identify chemical structures that could have good inhibitory effects on specific targets and have low toxicity (non-specific activity). Of special interest is the prediction of partition coefficient log P, which is an important measure used in identifying "druglikeness" according to Lipinski's rule of five.

While many quantitative structure activity relationship analyses involve the interactions of a family of molecules with an enzyme or receptor binding site, QSAR can also be used to study the interactions between the structural domains of proteins. Protein–protein interactions can be quantitatively analyzed for structural variations resulted from site-directed mutagenesis.

It is part of the machine learning method to reduce the risk for an SAR paradox, especially taking into account that only a finite amount of data is available (see also MVUE). In general, all QSAR problems can be divided into a coding and learning.

Drug design: Researchers have attempted for many years to develop drugs based on

QSAR. Easy access to computational resources was not available when these efforts began, so attempts consisted primarily of statistical correlations of structural descriptors with biological activities. However, as access to high-speed computers and graphics workstations became commonplace, this field has evolved into what is often termed rational drug design or computer-assisted drug design.

We will discuss the application of QSAR to drug design, some examples of which relied primarily on statistical correlation and some, on computer-based visualization and Modelling. An early example of QSAR in drug design involves a series of 1-(X-phenyl)-3,3-dialkyl triazenes.

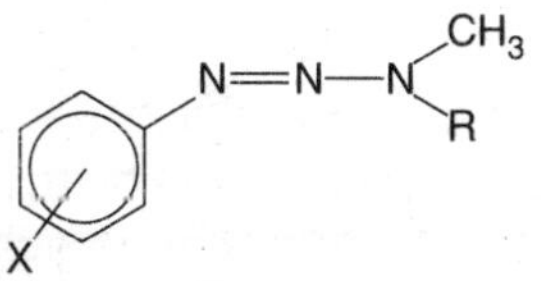

These compounds were of interest for their anti-tumor activity, but they also were mutagenic. QSAR was applied to understand how the structure might be modified to reduce the mutagenicity without significantly decreasing the anti-tumor activity. Mutagenic activity was evaluated in the Ames test, and from those data, the following QSAR was developed:

$$\log\left(\frac{1}{C}\right) = 1.09 \log P - 1.63\sigma^{+}3.06$$

where C is the molar concentration. From the equation, it is seen that factors that favor mutagenicity are increased lipophilicity and electron-donating substituents.

Studies of the anti-tumor activity were done against L1210 leukemia in mice. From the data, the following QSAR was developed:

$$\log\left(\frac{1}{C}\right) = 0.01 \log P - 0.042 (\log P)^2 - 0.31\sigma^{+}$$

$$-0.81MR + 0.39E_SR + 4.12$$

where C is the molar concentration of compound producing a 40% increase in life span of mice, MR is molar refractivity, which is a measure of molecular volume, and EsR is a steric parameter for the R group. Based on these equations, mutagenicity is more sensitive than anti-tumor activity to the electronic effects of the substituents. Thus, electron-withdrawing substituents were examined, as illustrated in the example below.

By substituting a sulfonamide group at the para position, the anti-tumor activity was reduced 1.2-fold, whereas the mutagenicity was reduced by about 400-fold.

	log (1/C)	
4–X substituent	Mutagenicity	Antitumorigenicity
–H	5.75	3.58
$-SO_2HN2$	3.15	3.48

Applicability Domain

As the use of (Q)SAR models for chemical risk management increases steadily and is also used for regulatory purposes (in the EU: registration, evaluation, authorisation and restriction of chemicals), it is of crucial importance to be able to assess the reliability of predictions. The chemical descriptor space spanned by a particular training set of chemicals is called Applicability Domain. It offers the opportunity to assess whether a compound can be reliably predicted.

Receptor Mapping

A receptor is a protein molecule, embedded in either the plasma membrane or the cytoplasm of a cell, to which a mobile signaling molecule may attach. A molecule which binds to a receptor is called a ligand, and may be a peptide or other small molecule, such as a neurotransmitter, a hormone, a pharmaceutical drug, or a toxin. When such binding occurs, the receptor undergoes a conformational change, which ordinarily initiates a cellular response. However, some ligands (e.g. antagonists) merely block receptors without inducing any response. Ligand-induced changes in receptors result in cellular changes which constitute the biological activity of the ligands.

The receptor–ligand complementarity is used to construct a receptor and which can be used to find out the unknown receptor. This process is called receptor mapping. In receptor mapping, bioactive conformations are derived for the active and inactive analogs and the volume of each set are superimposed. The active analog volume is subtracted from the inactive volume of analog and it is used to find the specific cavity where the active analogs fit. The inactive analogs are assumed to be inactive because of unfavorable steric interactions with the receptor. There are a number of programs available for the computation of molecular volume. Pseudo-electron density function calibrated to produce van der Waals radii, that have been mapped to a 3D grid to compute the union, intersection and subtraction of volume

The shapes and actions of receptors are studied by X-ray crystallography, computer modelling, and structure-function studies, which have advanced the understanding of drug action at the binding sites of receptors.

Depending on their functions and ligands, several types of receptors may be identified:

- Some receptor proteins are peripheral membrane proteins.
- Many hormone and neurotransmitter receptors are transmembrane proteins: transmembrane receptors are embedded in the phospholipid bilayer of cell membranes, that allow the activation of signal transduction pathways in response to the activation by the binding molecule, or ligand.
  - Metabotropic receptors are coupled to G proteins and affect the cell indirectly through enzymes which control ion channels.
  - Ionotropic receptors (also known as ligand-gated ion channels) contain a central pore which opens in response to the binding of ligand.
- Another major class of receptors are intracellular proteins such as those for steroid and intracrine peptide hormone receptors. These receptors often can enter the cell nucleus and modulate gene expression in response to the activation by the ligand.

Ligand Receptor Interaction

Ligand binding is an equilibrium process. Ligands bind to receptors and dissociate from them according to the law of mass action.

$$[\text{Ligand}].\,[\text{Receptor}] \underset{}{\overset{Kd}{\rightleftharpoons}} [\text{ligand–receptor complex}]$$

(the brackets stand for concentrations)

One measure of how well a molecule fits a receptor is the binding affinity, which is inversely related to the dissociation constant Kd. A good fit corresponds with high affinity and low Kd. The final biological response (e.g. second messenger cascade or muscle contraction), is only achieved after a significant number of receptors are activated.

The receptor-ligand affinity is greater than enzyme-substrate affinity. Whilst both interactions are specific and reversible, there is no chemical modification of the ligand as seen with the substrate upon binding to its enzyme.

The interaction of most ligands with their binding sites can be characterized in terms of a binding affinity. In general, high affinity ligand binding results from greater intermolecular force between the ligand and its receptor while low affinity ligand binding involves less intermolecular force between the ligand and its receptor. High affinity binding involves a longer residence time for the ligand at its receptor binding site than is the case for low affinity binding. High affinity binding of ligands to receptors is often physiologically important when some of the binding energy can be used to cause a conformational change in the receptor, resulting in altered behavior of an associated ion channel or enzyme.

Constitutive activity

A receptor which is capable of producing its biological response in the absence of a bound ligand is said to display "constitutive activity". The constitutive activity of receptors may be blocked by inverse agonist binding. Mutations in receptors that result in increased constitutive activity underlie some inherited diseases, such as precocious puberty (due to mutations in luteinizing hormone receptors) and hyperthyroidism (due to mutations in thyroid-stimulating hormone receptors).

Agonists versus antagonists

Not every ligand that binds to a receptor also activates the receptor. The following classes of ligands exist:

- (Full) agonists are able to activate the receptor and result in a maximal biological response. Most natural ligands are full agonists.
- Partial agonists do not activate receptors thoroughly, causing responses which are partial compared to those of full agonists.
- Antagonists bind to receptors but do not activate them. This results in receptor blockage, inhibiting the binding of other agonists.
- Inverse agonists reduce the activity of receptors by inhibiting their constitutive activity.

Pharmacophore

A pharmacophore was first defined by Paul Ehrlich in 1909 as "a molecular framework that carries (phoros) the essential features responsible for a drug's (=pharmacon's) biological activity". In 1977, this definition was updated by Peter Gund to "a set of structural features in a molecule that is recognized at a receptor site and is responsible for that molecule's biological activity". The IUPAC definition of a pharmacophore is "an ensemble of steric and electronic features that is necessary to ensure the optimal supramolecular interactions with a specific biological target and to trigger (or block) its biological response".

In modern computational chemistry, pharmacophores are used to define the essential features of one or more molecules with the same biological activity. A database of diverse chemical compounds can then be searched for more molecules which share the same features located a similar distance apart from each other.

Typical pharmacophore features are for where a molecule is hydrophobic, aromatic, a hydrogen bond acceptor, a hydrogen bond donor, a cation, or an anion. The features need to match different chemical groups with similar properties, in order to identify novel ligands. Ligands receptor interactions are typically "polar positive", "polar negative" or "hydrophobic". A well-defined pharmacophore model includes both hydrophobic volumes and hydrogen bond vectors.

Steps involved in pharmacophoric pattern determination:

1. Deriving preferred conformation of molecule via conformation search scheme.
2. Define the common groups in terms of specific atom types, general atom types(H accepting), charge property of atom and surfaces, functional group, etc.
3. Superposition and alignment of molecule at the specific point in a defined way.
4. Join the common site and calculate the distance by finding out the average sites of superposition in a least square fit calculation. A pharmacophore must have at least three points of connection and it can be represented geometrically as a triangle.

Pharmacophore modelling software

- Discovery Studio by Accelrys
- Phase by Schroedinger
- MOE by the Chemical Computing Group
- ICM-Chemist by MolSoft

11

Microarray Technology

A microarray is a tool for analyzing gene expression that consists of a small membrane or glass slide. With only a few exceptions; every cell of the body contains a full set of chromosomes and identical genes. Only a fraction of these genes are turned on, however, and it is the subset that is "expressed" that confers unique properties to each cell type. "Gene expression" is the term used to describe the transcription of the information contained within the DNA, the repository of genetic information, into messenger RNA (mRNA) molecules that are then translated into the proteins that perform most of the critical functions of cells. Scientists study the kinds and amounts of mRNA produced by a cell to learn which genes are expressed, which in turn provides insights into how the cell responds to its changing needs. Gene expression is a highly complex and tightly regulated process that allows a cell to respond dynamically both to environmental stimuli and to its own changing needs. This mechanism acts as both an "on/off" switch to control which genes are expressed in a cell as well as a "volume control" that increases or decreases the level of expression of particular genes as necessary. The use of a collection of distinct DNAs in arrays for expression profiling was first described in 1987, and the arrayed DNAs were used to identify genes whose expression is modulated by interferon. The use of miniaturized microarrays for gene expression profiling was first reported in 1995 and a complete eukaryotic genome (*Saccharomyces cerevisiae*) on a microarray was published in 1997.

Importance of Microarray

Microarrays are a significant advance both because they may contain a very large number of genes and because of their small size.

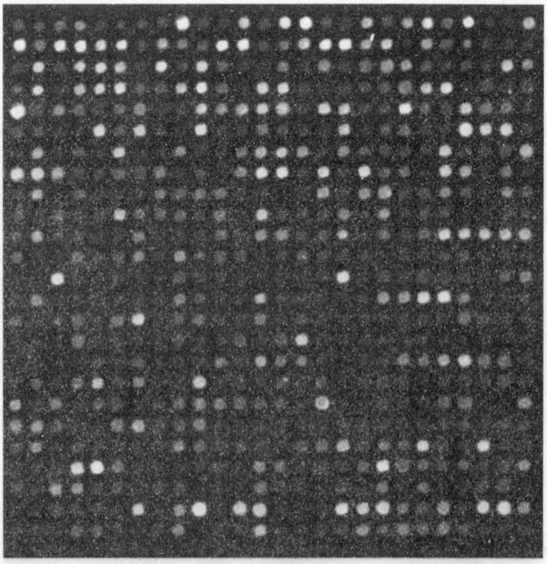

Microarrays are therefore useful when one wants to survey a large number of genes quickly or when the sample to be studied is small. Microarrays may be used to assay gene expression within a single sample or to compare gene expression in two different cell types or tissue samples, such as in healthy and diseased tissue. Because a microarray can be used to examine the expression of hundreds or thousands of genes at once, it promises to revolutionize the way scientists examine gene expression. This technology is still considered to be in its infancy; therefore, many initial studies using microarrays have represented simple surveys of gene expression profiles in a variety of cell types. Nevertheless, these studies represent an important and necessary first step in our understanding and cataloging of the human genome. As more information accumulates, scientists will be able to use microarray to ask increasingly complex questions and perform more intricate experiments. With new advances, researchers will be able to infer probable functions of new genes based on similarities in expression patterns with those of known genes. Ultimately, these studies promise to expand the size of existing gene families, reveal new patterns of coordinated gene expression across gene families, and uncover entirely new categories of genes.

Furthermore, because the product of any one gene usually interacts with those of many others, our understanding of how these genes coordinate will become clearer through such analyses, and precise knowledge of these interrelationships will emerge. The use of microarrays may also speed the identification of genes involved in the development of various diseases by enabling scientists to examine a much larger number of genes. This technology will also aid the examination of the integration of gene expression and function at the cellular level, revealing how multiple gene products work together to produce physical and chemical responses to both static and changing cellular needs. A basic difference between microarray data analysis and much traditional biomedical research is the dimensionality of the data. A large clinical study might collect 100 data items per patient for thousands of patients. A medium-size microarray study will obtain many thousands of numbers per sample for perhaps a hundred samples. Many analysis techniques treat each sample as a single point in a space with thousands of dimensions, then attempt by various techniques to reduce the dimensionality of the data to something humans can visualize.

DNA Microarrays: The Technical Foundations

Two recent complementary advances, one in knowledge and one in technology, are greatly facilitating the study of gene expression and the discovery of the roles played by specific genes in the development of disease. As a result of the Human Genome Project, there has been an explosion in the amount of information available about the DNA sequence of the human genome. Consequently, researchers have identified a large number of novel genes within these previously unknown sequences. The challenge currently facing scientists is to find a way to organize and catalog this vast amount of information into a usable form. Only after the functions of the new genes are discovered will the full impact of the Human Genome Project be realized. The second advance may facilitate the identification and classification of this DNA sequence information and the assignment of functions to these new genes: the emergence of DNA microarray technology. A microarray works by exploiting the ability of a given mRNA molecule to bind specifically to, or hybridize to, the DNA template from which it originated. By using an array containing many

DNA samples, scientists can determine, in a single experiment, the expression levels of hundreds or thousands of genes within a cell by measuring the amount of mRNA bound to each site on the array.

With the aid of a computer, the amount of mRNA bound to the spots on the microarray is precisely measured, generating a profile of gene expression in the cell.

What Exactly is a DNA Microarray

DNA microarrays are small, solid supports onto which the sequences from thousands of different genes are immobilized, or attached, at fixed locations. The supports themselves are usually glass microscope slides, the size of two side-by-side pinky fingers, but can also be silicon chips or nylon membranes. The DNA is printed, spotted, or actually synthesized directly on to the support. The American Heritage Dictionary defines "array" as "to place in an orderly arrangement". It is important that the gene sequences in a microarray are attached to their support in an orderly or fixed way, because a researcher uses the location of each spot in the array to identify a particular gene sequence. The spots themselves can be DNA, cDNA, or oligonucleotides.

Steps of Microarray Experiments

One might ask, how does a scientist extract information about a disease condition from a dime-sized glass or silicon chip containing thousands of individual gene sequences? The whole process is based on hybridization probing, a technique that uses fluorescently labeled nucleic acid molecules as "mobile probes" to identify complementary molecules, and sequences that are able to base-pair with one another. Each single-stranded DNA fragment is made up of four different nucleotides, adenine (A), thymine (T), guanine (G), and cytosine (C) that are linked end to end. Adenine is the complement of, or will always pair with, thymine, and guanine is the complement of cytosine. Therefore, the complementary sequence to G-T-C-C-T-A will be C-A-G-G-A-T. When two complementary sequences find each other, such as the immobilized target DNA and the mobile probe DNA, cDNA, or mRNA, they will lock together, or hybridize.

Now, consider two cells: cell type 1, a healthy cell, and cell type 2, a diseased cell. Both contain an identical set of four genes, A, B, C, and D. Scientists are interested in determining the expression profile of these four genes in the two cell types. To do this, scientists isolate mRNA from each cell type and use this mRNA as templates to generate cDNA with a "fluorescent tag" attached. Different tags (red and green) are used so that the samples can be differentiated in subsequent steps. The two labeled samples are then mixed and incubated with a microarray containing the immobilized genes A, B, C, and D. The labeled molecules bind to the sites on the array corresponding to the genes expressed in each cell. After this hybridization step is complete, a researcher will place the microarray in a "reader" or "scanner" that consists of some lasers, a special microscope, and a camera. The fluorescent tags are excited by the laser, and the microscope and camera work together to create a digital image of the array. These data are then stored in a computer, and a special program is used either to calculate the red-to-green fluorescence ratio or to subtract out background data for each microarray spot by analyzing the digital image of the array. If calculating ratios, the program then creates a table that contains the ratios of the intensity of red-to-green fluorescence for every spot on the array. For example, using the scenario outlined above, the computer may conclude that both cell types express gene A at the same level, that cell 1 expresses more of gene B, that cell 2 expresses more of gene

C, and that neither cell expresses gene D. But remember, this is a simple example used to demonstrate key points in experimental design. Some microarray experiments can contain up to 30,000 target spots. Therefore, the data generated from a single array can mount up quickly.

Irrespective of the specific technology employed, a microarray experiment consists of three basic steps: sample preparation and labeling, sample hybridization and washing, and microarray image scanning and processing. In the following section, the cDNA microarray will serve as a basis for a general discussion of these steps. In general, experiments based on other platforms, such as the Affymetrix GeneChip, follow similar principles unless specifically stated.

Sample Preparation and Labeling

Sample preparation involves extracting and purifying the mRNAs from the tissue of interest. Due to a number of challenges, this procedure can be quite variable. First, the target mRNA typically accounts for only a small fraction (less than 3%) of all mRNA in a cell. Second, it could be very difficult to isolate mRNA specific to the study from a heterogeneous range of cells (e.g. diseased tissue contains a mixture of normal tissue, inflammatory cells, necrotic tissue, and, in cancer samples, areas of different grades). Finally, captured mRNA degrades very quickly. To address this rapid degradation, the mRNA is usually reverse-transcribed into more stable cDNA (for cDNA microarrays) immediately after extraction. To allow detection of which cDNAs are bound to the microarray, the sample goes through a platform-dependent labeling process. For Affymetrix platform, a biotin-labeled complementary RNA is constructed for hybridizing to the GeneChip. The protocols are very carefully defined by Affymetrix to ensure that every Affymetrix laboratory follows identical steps. Experimental results obtained in different Affymetrix laboratories should therefore be reliably comparable. Detection of DNA on cDNA microarrays had previously been performed using radioactively-labeled DNA, but it is now more common to use dyes which fluoresce when exposed to a specific wavelength of light. In the most common experiments, two samples are hybridized to arrays, each labeled with Cy3 or Cy5 dyes, which are excited by green and red lasers, respectively. This results in a two-channel cDNA microarray and allows the simultaneous measurement of both samples. In the future, it is possible that more than two labeled samples may be used, producing a multichannel cDNA microarray.

Hybridization

Hybridization is the step in which the DNA probes on the microarrays and the labeled DNA (or RNA) target forms heteroduplexes according to the Watson-Crick base-pairing rule. The essential principle here is that a single-stranded DNA molecule will bind to another single-stranded DNA molecule with a precisely matching sequence with much higher affinity than that to an imperfectly matching sequence. In reality, however, hybridization is a complex process and a DNA segment may also bind well to a sequence similar but not identical to its complementary target, a phenomenon called cross-hybridization. This is influenced by many conditions, including temperature, humidity, salt concentration, formamide concentration, target solution volume, and hybridization operator. Hybridization may be performed either manually or by a robot. Robotic hybridization provides much better control over the temperatures of the target and slide.

The consistent use of a single hybridization station also reduces the variability which arises from multiple hybridizations and various operators. After hybridization, the microarray is removed from the chamber or station and is then washed to eliminate any excess labeled sample so that only the DNA complementary to the probes remains bound on the array. Finally, the microarray is dried using a centrifuge or by blowing clean compressed air.

Image Scanning

After the completion of hybridization, the surface of the hybridized array is scanned to produce a microarray image. As previously mentioned, samples are labeled with biotin or fluorescent dyes that emit detectable light when stimulated by a laser. The emitted light is captured by the photo-multiplier tube (PMT) in a scanner, and the intensity is recorded. Most scanners contain one or more lasers that are focused onto the array (for two-channel microarrays, the scanner uses two lasers).

Although the scanner is only intended to detect light emitted by the target DNA strands which are bound to their complementary probes, it also will capture incidental light from various other sources. These other sources may include labeled DNA sample which has hybridized non-specifically to the glass slide, residual (unwashed) labeled sample which has adhered to the slide, various chemicals used in processing the slide, and even the slide itself. This incidentally-captured light is called background. The scanned output of an Affymetrix chip is usually a monochrome image. With two-channel microarrays, the output is a pair of monochrome images. Each image is from one of the lasers in the scanner. The two monochrome images are combined to create the false color images of microarrays. Both monochrome and two-color images are usually stored in the tagged image file format (TIFF).

Image Processing

The microarray image generated by the scanner forms the raw data of a microarray experiment. Prior to data analysis, the image must be converted from the TIFF format into the numerical information that quantifies gene expression. The manner in which this is accomplished will have a major impact on the quality of the resulting data and the success of further analysis. For in situ synthesized microarrays, both Affymetrix and Agilent have integrated tailored image-processing algorithms into their software packages, allowing end-users to directly generate quantified microarray data.

For cDNA microarrays, the images consist of spots arranged in regular grid-like patterns. The processing of these images consists of four basic steps:

1. *Spot identification:* The process of spot identification involves locating the position of individual signal spots in an image and estimating their size.
2. *Imagesegmentation:* This process involves decomposing an image into a set of non-overlapping regions. Specifically, this step involves differentiation of those pixels which form the spot and should be included in the calculation of the signal from those pixels which are background or noise and should be eliminated.
3. *Spot quantification:* After the pixels belonging to the signal and the background have been distinguished, this process involves calculating the intensity for each spot. Here, pixel intensity values are combined into a unique value representing the expression level of a gene deposited in a given spot.
4. *Spot quality assessment:* This process involves calculating some quality-control measures which evaluate the quality of both the entire array and the individual

spots on the array. These measures can assist human inspectors in determination of the data reliability and identification of those spots with questionable quality values.

Methodology

DNA microarrays are created by spotting every gene in a genome on to a glass microscope slide (Fig. 11.1).

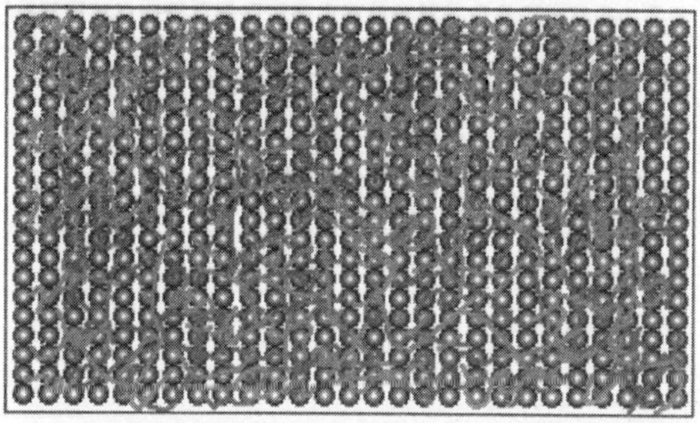

RNA is isolated from two populations of cells. Two populations of RNA are converted to two colors (red and green) of cDNA.

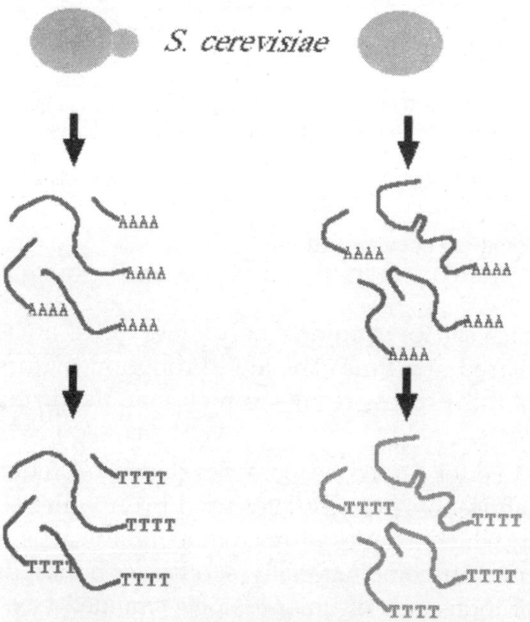

cDNAs are hybridized with the microarray

Spots that bind to both cDNAs appear yellow

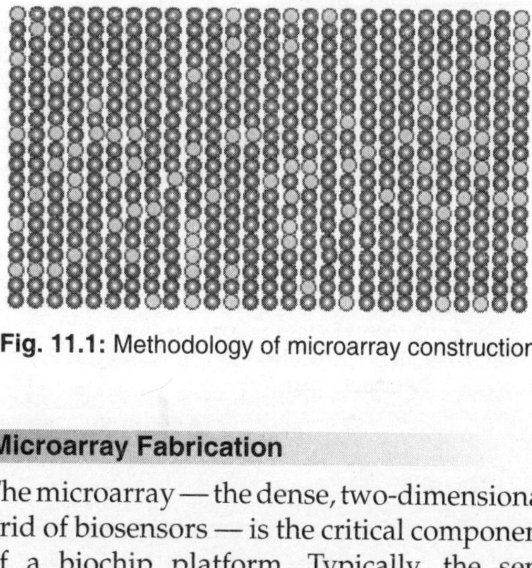

Fig. 11.1: Methodology of microarray construction

Microarray Fabrication

The microarray — the dense, two-dimensional grid of biosensors — is the critical component of a biochip platform. Typically, the sensors are deposited on a flat substrate, which may either be passive (e.g. silicon or glass) or active, the latter consisting of integrated electronics or micromechanical devices that perform or assist signal transduction. Surface chemistry is used to covalently bind the sensor molecules to the substrate medium. The fabrication of microarrays is non-trivial and

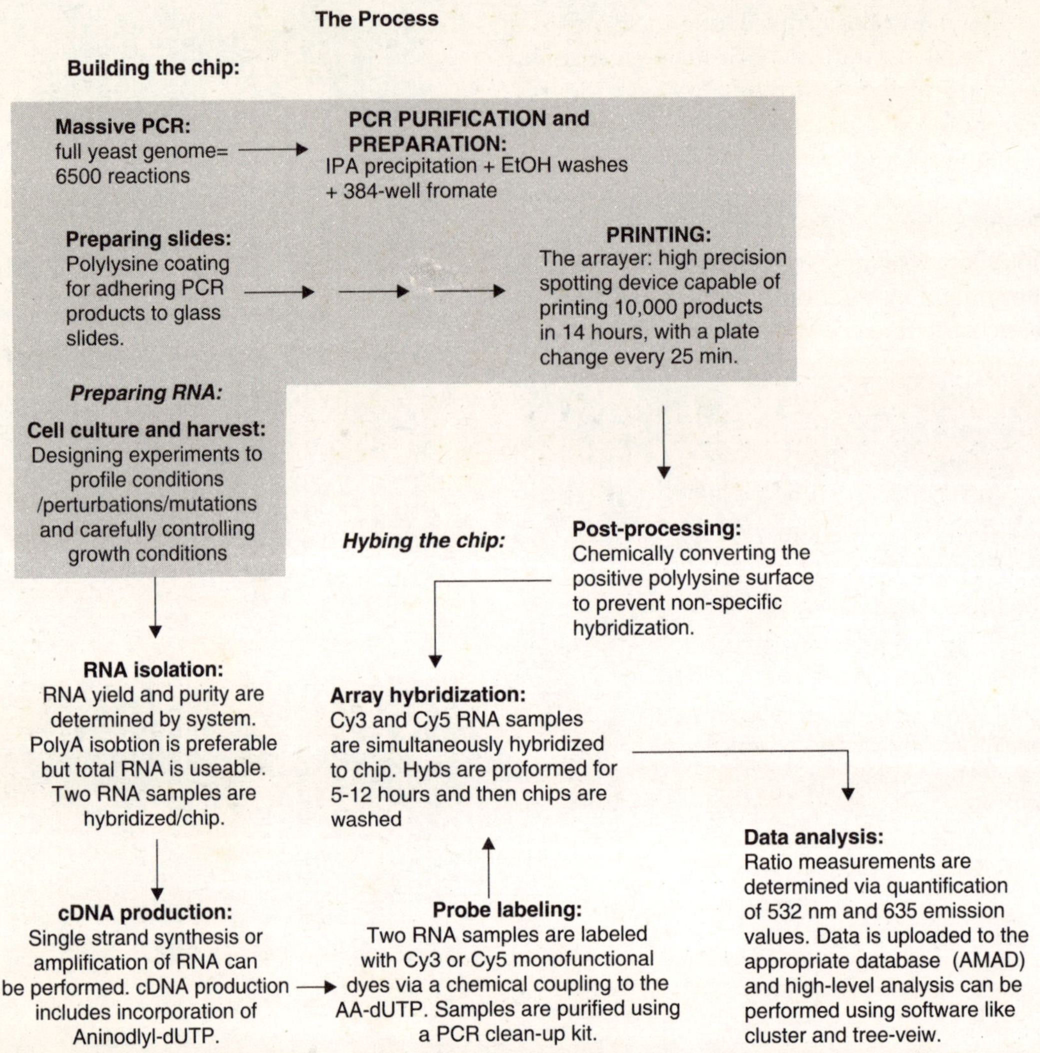

Fig. 11.2: Flow chart of microarray construction

is a major economic and technological hurdle that may ultimately decide the success of future biochip platforms. The primary manufacturing challenge is the process of placing each sensor at a specific position (typically on a Cartesian grid) on the substrate. Various means exist to achieve the placement, but typically robotic micro-pipetting or micro-printing systems are used to place tiny spots of sensor material on the chip surface. Because each sensor is unique, only a few spots can be placed at a time. The low-throughput nature of this process results in high manufacturing costs.

Fodor and colleagues developed a unique fabrication process (later used by Affymetrix) in which a series of microlithography steps is used to combinatorially synthesize hundreds of thousands of unique, single-stranded DNA sensors on a substrate one nucleotide at a time.

One lithography step is needed per base type; thus, a total of four steps is required per nucleotide level. Although this technique is very powerful in that many sensors can be created simultaneously, it is currently only feasible for creating short DNA strands (15–25 nucleotides). Reliability and cost factors limit the number of photolithography steps that can be done. Furthermore, light-directed combinatorial synthesis techniques are not currently possible for proteins or other sensing molecules.

As noted above, most microarrays consist of a Cartesian grid of sensors. This approach is used chiefly to map or "encode" the coordinate of each sensor to its function. Sensors in these arrays typically use a universal signaling technique (e.g. fluorescence), thus making coordinates their only identifying feature. These arrays must be made using a serial process (i.e. requiring multiple, sequential steps) to ensure that each sensor is placed at the correct position."Random" fabrication, in which the sensors are placed at arbitrary positions on the chip, is an alternative to the serial method. The tedious and expensive positioning process is not required, enabling the use of parallelized self-assembly techniques. In this approach, large batches of identical sensors can be produced; sensors from each batch are then combined and assembled into an array. A non-coordinate-based encoding scheme must be used to identify each sensor.

Standardization

Microarray data is difficult to exchange due to the lack of standardization in arrays. This presents an interoperability problem in bioinformatics. Various grassroots open-source projects are trying to ease the exchange and analysis of data produced with non-proprietary chips: For example, the "minimum information about a microarray experiment" (MIAME) checklist helps define the level of detail that should exist and is being adopted by many journals as a requirement for the submission of papers incorporating microarray results. But MIAME does not describe the format for the information, so while many formats can support the MIAME requirements, as of 2007 no format permits verification of complete semantic compliance. The "MicroArray Quality Control (MAQC) Project" is being conducted by the US Food and Drug Administration (FDA) to develop standards and quality control metrics which will eventually allow the use of microarray data in drug discovery, clinical practice and regulatory decision-making. The microarray and Gene Expression Data (MGED) group is working on the standardization of the representation of gene expression data and relevant annotations.

Biochips

The basic difference between biochips and microarray lies in the design. The term "biochip" was derived from the term "computer chip".

A biochip is a collection of miniaturized test sites (microarrays) arranged on a solid substrate that permits many tests to be performed at the same time in order to achieve higher output and speed. Biochips can also be used to perform techniques such as electrophoresis or PCR using microfluidics technology.

The development of biochips is a major thrust of the rapidly growing biotechnology

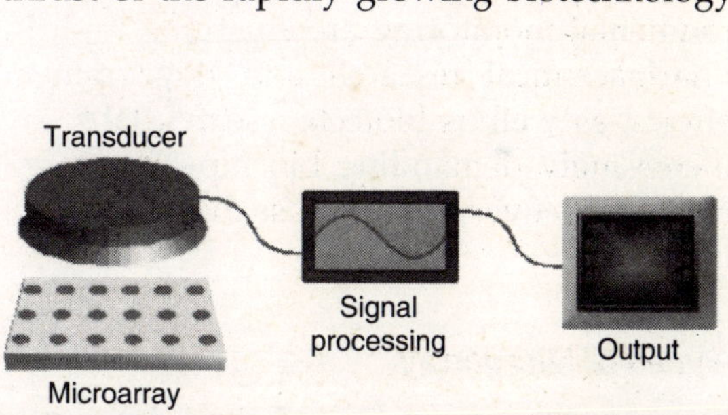

Fig. 11.3: Biochips are a platform that require, in addition to microarray technology, transduction and signal processing technologies to output the results of sensing experiments

industry, which encompasses a very diverse range of research efforts including genomics, proteomics, and pharmaceuticals, among other activities. Advances in these areas are giving scientists new methods for unraveling the complex biochemical processes occurring inside cells, with the larger goal of understanding and treating human diseases. At the same time, the semiconductor industry has been steadily perfecting the science of micro-miniaturization. The merging of these two fields in recent years has enabled biotechnologists to begin packing their traditionally bulky sensing tools into smaller and smaller spaces, onto so-called biochips. These chips are essentially miniaturized laboratories that can perform hundreds or thousands of simultaneous biochemical reactions. Biochips enable researchers to quickly screen large numbers of biological analytes for a variety of purposes, from disease diagnosis to detection of bioterrorism agents.

Market Demand

Current demand for biochips is driven by basic research, which involves genome mapping and sequencing laboratories and drug discovery applications. Currently, in Europe, pharmaceutical research is receiving the most attention, especially drug discovery applications of the biochip technology. Future markets, such as diagnostic testing, and environmental monitoring are emerging rapidly. Pharmaceutical research and development efforts, as well as biotech research labs, are increasingly demanding biochip to be used in drug discovery and disease discovery applications.

Types of Microarray

Microarrays are classified based on the molecules used as probes and the sample molecules. They vary in the interaction between immobilized and the labeled molecules. Following are the different types of microarrays:

1. DNA microarray—this harnesses the ability of nucleic acid with complementary sequences to hybridize to each other under suitable conditions. Here the probes are different sequences of nucleic acids and they are used to interrogate the complex mixture of labeled nucleic acids for example cDNA's in transcriptomics experiments.
2. Protein microarrays—here the probes are proteins. Proteins, substrates, nucleic acids constitute the labeled molecules. This is used to study receptor-ligand, enzyme-substrate reactions.
3. Carbohydrate microarray (glycochip)—here the probes are carbohydrate motifs. These are used to study the protein-carbohydrate interactions. This can be used to identify the organism with organism specific protein-carbohydrate interactions.
4. Small molecule arrays—here different small molecules are listed for their affinity for certain protein motifs. These are invaluable tools in discovering a lead molecule as a drug.
5. RNA microarray—it consists of RNAs that fold into intricate 3D structures and selectively recognize a wide range of targets, a function that is similar to antibodies. Therefore, advanced RNA biochips should be able to be used to detect almost anything that RNA can be made to bind. RNA biochip can be used to see many different targets like drugs, toxins and metabolites, as well as both proteins and nucleic acids.
6. Tissue microarrays (also TMAs)—consist of paraffin blocks in which up to 1000 separate tissue cores are assembled in array fashion to allow simultaneous histological analysis. The major limitations in molecular clinical analysis of

tissues include the cumbersome nature of procedures, limited availability of diagnostic reagents and limited patient sample size.

In Brief: Microarray Applications

- Gene expression studies.
- Gene function for cell state change in various conditions (clustering, classification).
- Disease diagnosis (classification).
- Inferring regulatory networks.
- Pathogen analysis.
- Drug discovery.
- Identify appropriate molecular targets for therapeutic intervention.
- Monitor changes in gene expression in response to drug treatments.
- Targeted drug treatment.
- Agro industries.
- Water quality management.
- Pharmacogenomics.
- Detection of biological warfare.

DNA MICROARRAY

A DNA microarray is a multiplex technology used in molecular biology and in medicine. DNA microarrays are miniaturized solid support on to which the sequence from thousands of different genes are immobilized or attached at fixed location. The support may be glass slide, silicon chip or nylon membrane.

In standard microarrays, the probes are attached to a solid surface by a covalent bond to a chemical matrix (via epoxy-silane, amino-silane, lysine, polyacrylamide or others). The solid surface can be glass or a silicon chip, in which case they are commonly known as gene chip or colloquially Affy chip when an Affymetrix chip is used. Other microarray platforms, such as Illumina, use microscopic beads, instead of the large solid support. DNA arrays are different from other types of microarray only in that they either measure DNA or use DNA as part of its detection system.

History of DNA Microarray

1987: Patent filed on sequencing by Hybridisation (SBH) R. Drmanac, Belgrade Univ. ® Argonne National Laboratory ® HySeq.

1988.1991: Several groups published reports on SBH , E. Southern, Oxford University (Oxford Gene Technology), A.Mirzabekov, Engelhard Institute, Moscow ® Agronne National Laboratory, S. Fodor, Affymetrix , W. Bains, Bath University.

1989: European patent granted to Southern on. Oligonucleotide arrays as a testing platform, Affymetrix genechip prototype was developed.

1991: Photolithographic printing (Affymetrix).

1993: US patent on SBH granted to HySeq

1994: First cDNA collections are developed at Stanford, First commercial Affymetrix genechip was developed, First commercial scanner was developed.

1995: Quantitative monitoring of gene expression patterns with a complementary DNA microarray.

1996: Commercialization of arrays (Affymetrix).

1997: Genome- wide expression monitoring in S. cerevisiae (yeast).

1997.1998: HySeq accuses Affymetrix of patent infringement we are not doing sequencing, but mutation detection.

1998: Court case between Southern and seven chip companies (Affymetrix, HySeq, Hoffman La Roche, Abbot, etc.).

1998: US patent to Incyte (Synteni) on technology to print microarrays with density higher than 100 polynucleotide per cm^2.

1998.1999: Affymetrix and Incyte (and others) accuse each other of patent infringements.

2000: Portraits/ Signatures of cancer.

2003: Introduction into clinical practices.

2004: Whole human genome on one microarray.

Uses of DNA Microarray

DNA microarrays can be used to detect DNA (as in comparative genomic hybridization), or detect RNA (most commonly as cDNA after reverse transcription) that may or may not be translated into proteins. The process of measuring gene expression via cDNA is called expression analysis or expression profiling. DNA microarrays can be used to detect single nucleotide polymorphisms (SNPs), in genotyping or in resequencing mutant genomes.

Since an array can contain tens of thousands of probes, a microarray experiment can

Table 11.1: Uses of DNA microarray

Technology or Application	*Synopsis*
Gene expression profiling	In an mRNA or gene expression profiling experiment, the expression levels of thousands of genes are simultaneously monitored to study the effects of certain treatments, diseases, and developmental stages on gene expression. For example, microarray-based gene expression profiling can be used to identify genes whose expression is changed in response to pathogens or other organisms by comparing gene expression in infected to that in uninfected cells or tissues.
Comparative genomic hybridization	Assessing genome content in different cells or closely related organisms.
GeneID	Small microarrays to check IDs of organisms in food and feed (like GMO), mycoplasms in cell culture, or pathogens for disease detection, mostly combining PCR and microarray technology.
Chromatin immunoprecipitation on Chip	DNA sequences bound to a particular protein can be isolated by immunoprecipitating that protein (ChIP), these fragments can be then hybridized to a microarray (such as a tiling array) allowing the determination of protein binding site occupancy throughout the genome. Example protein to immunoprecipitate are histone modifications ($H_3K_{27}me_3$, $H_3K_4me_2$, $H3K_9me_3$, etc), Polycomb-group protein (PRC_2:$Suz1_2$, PRC1:YY1) and trithorax-group protein (Ash1) to study the epigenetic landscape or RNA Polymerase II to study the transcription landscape.
SNP detection	Identifying single nucleotide polymorphism among alleles within or between populations. Several applications of microarrays make use of SNP detection, including Genotyping, forensic analysis, measuring predisposition to disease, identifying drug-candidates, evaluating germline mutations in individuals or somatic mutations in cancers, assessing loss of heterozygosity, or genetic linkage analysis.
Alternative splicing detection	An exon junction array design uses probes specific to the expected or potential splice sites of predicted exons for a gene. It is of intermediate density, or coverage, to a typical gene expression array (with 1–3 probes per gene) and a genomic tiling array (with hundreds or thousands of probes per gene). It is used to assay the expression of alternative splice forms of a gene. Exon arrays have a different design, employing probes designed to detect each individual exon for known or predicted genes, and can be used for detecting different splicing isoforms.

Contd...

Contd...

Technology or Application	*Synopsis*
Fusion genes microarray	A fusion gene microarray can detect fusion transcripts, e.g. from cancer specimens. The principle behind this is building on the alternative splicing micorrays. The oligo design strategy enables combined measurements of chimeric transcript junctions with exon-wise measurements of individual fusion partners.
Tiling array	Genome tiling arrays consist of overlapping probes designed to densely represent a genomic region of interest, sometimes as large as an entire human chromosome. The purpose is to empirically detect expression of transcripts or alternatively splice forms which may not have been previously known or predicted.

accomplish that many genetic tests in parallel. Therefore, arrays have dramatically accelerated many types of investigation.

Fabrication: Process of fabrication of microarray involves following steps:

- Preparation of sample.
- Construction of array.
- Preparation of probe.
- Hybridization.
- Detection of result.
- Normalization.
- Data analysis using bioinformatics tool.
- Submitting the results to microarray database for future use.
  - Preparation of sample:
  - Required genome is extracted from the source and purified
  - Polymerase chain reaction used for amplifying the isolated DNA.
  - Construction of array.
  - DNA sequence is printed on to the microscope slides robotically, in the specified position.

The three main methods for constructing the arrays are:

a. Spotting of DNA fragments directly on to the slide
   - Microspotting
   - Ink-jetting
b. Piezoelectric printing
c. Photolithography

a. Spotting of DNA fragments directly on to the slide

The use of prefabricated oligonucleotides greatly simplifies the construction of microarrays. Conventional methods can be used to produce the sequences, and these can then be printed directly on to the microscope slide (which is first overlaid with a coating that is positively charged). The cDNA or oligonucleotide material can be deposited straight from a reagent tray (where it is suspended in a denaturing solution) onto the glass surface by a print head containing microspotting pins or micropipettes.

Microspotting: Printhead will carry the sample and it touches the surface of glass slide and delivers sample in fixed location, and moves to next place.

Ink-jetting: Printhead will carry the sample and its tip will not touch the surface of slide. Without touching, it will deliver the sample on predetermined position

Preparation of the probes: Fluorescent probes must be prepared, and these hybridise to the microarray. These are prepared from messenger RNA from the cells or tissues of interest. Extraction of the mRNA is made easier by the identification of a key property by which the mRNA can be isolated.

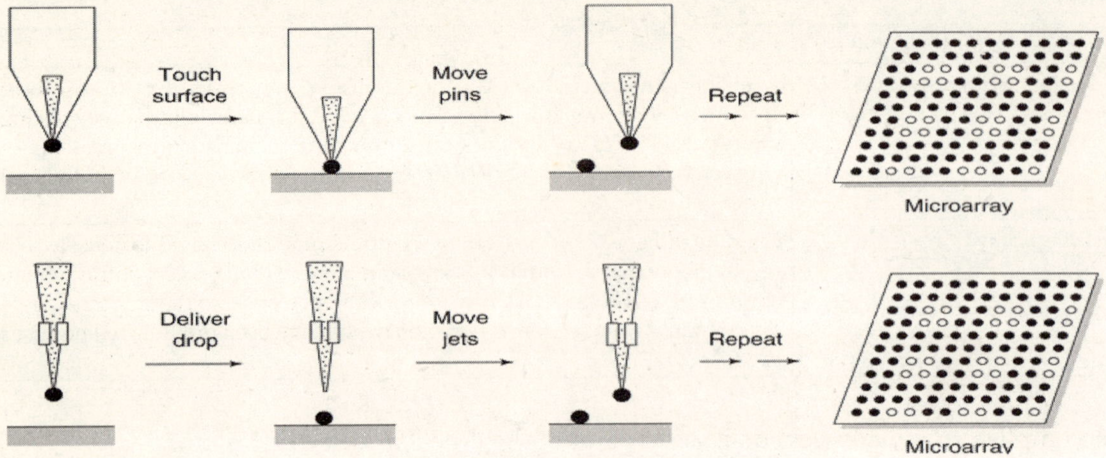

Fig. 11.4: Inkjet printing

After the mRNA has been extracted from the cells or tissues under study, it is converted into DNA by the use of the reverse transcriptase enzyme. During this reaction, the DNA is labeled by the incorporation of fluorescent or radioactive nucleotides into the DNA. The two samples are labelled using two different fluorescent dyes, red or green. The most common dyes in use are Cy3 and Cy5. This labeled DNA is then hybridised to the microarray slide. Using mixtures that are differentially labeled allows the ratio of fluorescence to be measured, therefore avoiding most of the problems of hybridisation kinetics. {Cy-Cyanin}

Hybridisation: Hybridisation is the reaction that occurs between the fluorescent probes and the DNA on the microarray.

The hybridisation conditions also depend on the application of the array:

- Detecting mutations requires high hybridisation stringencies: lower salt concentration and higher temperature, over short time periods (hours)
- Expression monitoring requires lower stringencies, to ensure low-copy number sequences anneal: overnight hybridisations with lower temperatures and higher salt concentrations.

Detection of result: After hybridisation, the microarray is scanned using a laser confocal scanning microscope, which illuminates each spot of DNA and separately measures the fluorescence for each dye. This produces data to determine the ratio, and in turn the relative abundance of the sequences of each specific gene in the messenger RNA or DNA samples. The hybridisation pattern can then be used to identify the genes that are expressed differently in the tissues or cells. After hybridisation and the readout of the expression levels obtained from the array, the vast quantity of data that is collected has to be stored and saved.

This then enables processing and analysis—both statistical and biological—to be carried out on the data. This is vital, because without significant image processing, the information content of a microarray is very limited.

Red intensity = R - R

fg = foreground

bg = background

Green intensity = G - G
M= intensity = log (R/G)
= log R – log G
Average intensity = log ({RG} ^1/2)
= log (R*G)/2
= (log R+ log G)/2

Normalization: Image analysis of microarray measures the foreground and background values in both red and green channel and a range of quality matrix for each spot on the array. Normalization is a process which utilizes this information to adjust for systematic differences in the relative intensity of each channel. The need for normalisation arises when two identical mRNA samples are labeled with different dyes and hybridised to the same slide.

DNA Array Readout Methods

Once a DNA array experiment has been designed and executed, the data must be extracted and analyzed. That is, the signal from each address on the array must be measured and some methods for determining and subtracting the background signal must be employed. However, because there are many different DNA array formats and platforms, and because hybridization signals can be generated with fluorescent- or radioactive-labeled targets, no single DNA array readout device is suitable for all purposes. Furthermore, many instruments with different advantages and disadvantages for different types of array formats are available. Therefore, since accurate data acquisition is a critical step of any array experiment, careful attention must be paid to the selection of data acquisition equipment.

Reading data from a fluorescent signal

All arrays that emit a fluorescent signal must be read with an instrument that provides a fluorescence excitation energy source and an efficient detector for the light emitted from the fluorophore incorporated into the target. Currently, the fluorophores most commonly used for incorporation into cDNA targets are Cy3 and Cy5. These are cyanine dyes commercially available as dUTP or dCTP conjugates. Cy3 is an orange dye with a light absorption maximum at 550 nm and an emission maximum at 581 nm. Cy5 is a far-red dye with a light absorption maximum at 649 nm and an emission maximum at 670 nm. All commercial scanners for reading fluorescent array signals use lasers and filters that provide an excitation light source at a single wavelength near the absorption maxima of these and other commonly used dyes. Lasers eliminate pollution of fluorescence emissions at nearby wavelengths that would occur with a light source of multiple wavelengths. Defined laser emission wavelengths and filters also permit the use and efficient detection of fluorescent emissions from two or more fluorophores in a single experiment.

This permits the common practice of combining cDNA targets prepared from a reference and an experimental condition, one containing Cy3-labeled and the other Cy5-labeled nucleotides, for analysis on a single array. Of course, instruments for this application must be equipped with multiple laser sources, one for each fluorophore used.

In addition to providing an appropriate light source for fluorophore excitation, attention must be paid to the efficiency and accuracy of fluorescence emission measurements. Efficiency is an issue because the amount of light emitted from a fluorophore is generally orders of magnitudes (as much as 1000000-fold) weaker than the intensity of the light required to excite the fluorophore. This problem is further exacerbated by the fact that fluorophores emit light in all directions, not just toward the light detector system. Thus, some optical method to collect as much of the emitted light and eliminate as much of the excitation light as possible is required. This is accomplished with an objective lens that captures the emitted light and focuses it toward the detector. However, it is not possible to collect this light simultaneously from all directions. The best that can be accomplished is to collect that light that is emitted from the hemisphere of the fluorescence directed toward the detector. The efficiency of an objective lens in accomplishing this task is expressed as its numerical aperture. If it were possible to collect all of the light in a hemisphere, the objective would have a numerical aperture of 1.0. In reality, objective lenses in array readers have numerical apertures ranging from 0.5 to 0.9. Obviously, the numerical aperture of the objective lens directly affects an instrument's sensitivity. Therefore, since instrument manufacturers often do not list this information in their product description, it is something that a purchaser should ask for.

All organic molecules fluoresce. Any contamination on an array surface (commonly glass slides), including the coated surface of the slide itself, will produce background fluorescence that can seriously compromise the signal-to-noise ratio. The answer to this problem has been addressed by limiting the focus of the laser to a small field of view in three-dimensional space centered in the DNA sample on the array. This is accomplished by a confocal method comprising two lenses in series producing two reciprocal focal points, one focused on the sample and the other focused at a pinhole for transmission of the light signal to the detector. The beauty of this system is that the pinhole permits only the light from a very narrow depth of focus of the objective lens to be passed through to the detector. This allows very fine discrimination of noise and signal.

Confocal systems image only a small point (pixel) in three-dimensional space. This is achieved by employing a laser as an illumination source and a small aperture in front of

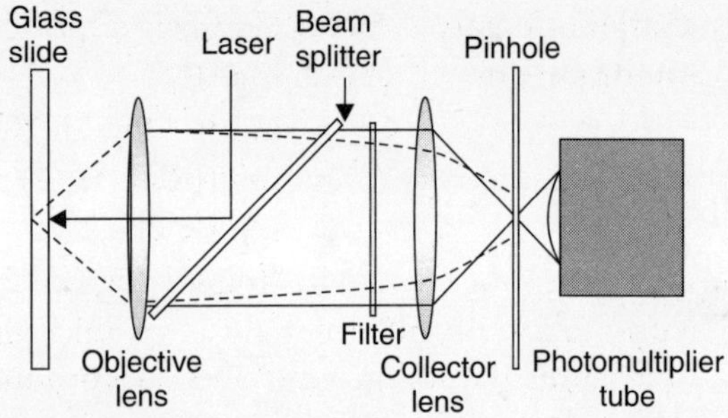

Fig. 11.5: The principle of confocal scanning laser microscopy

the detector, which usually is a photomultiplier tube. A schematic of a simplified confocal optical path is shown above. The laser beam enters from above and is reflected by the beam splitter (a dichroic or multichroic filter that reflects light of short wavelengths and transmits light of longer wavelengths). The laser beam is focused to a spot on the surface of the glass slide by an objective lens and excites fluorophores in the focal plane of the objective. The fluorescent emissions (and the reflected laser beam rays) are collected by the objective and collimated into parallel rays (solid lines). Most of the reflected laser rays (~95%) are reflected back toward the laser source by the beam splitter. The remaining laser rays are excluded by a downstream filter specific for the emission wavelength of the fluorophore. The parallel rays of the fluorescent light received from the focal plane of the objective lens are focused on a pinhole in a detector plate and passed through to the photomultiplier tube. The ray paths indicated with dashed lines shows how light from out-of-focus objects such as reflections and emissions from the second surface of the glass slide are eliminated by the design of the optical system. This out-of-focus fluorescent light takes a different path through the objective lens, the beam splitter, and the emission filter. As a result, it is not focused into the detector pinhole and, therefore only a small portion of this polluted light is passed on to the photomultiplier tube. Since confocal microscopy images only a small point in three-dimensional space proportional to the aperture of the pinhole in the detector plate, complete images must be obtained by scanning and digital reconstruction.

The depth of focus for most commercial laser-based array readers is 3 µm, less than half the thickness of the DNA sample on the slide surface. However, since the diameters of the spots printed on glass slide DNA arrays range from 25–100 µm, this means that at this 3 µm (1 pixel) resolution slides must be scanned to collect all of the data even from a single spot. At this resolution, even the highest density arrays can be accurately scanned. Scanning at this resolution, however, requires several minutes. For this reason, manufacturers of confocal laser scanners provide the option to scan at lower resolutions ranging down to 50 µm. This is useful for rapid pre-scanning of arrays for sensitivity and scanning alignment adjustments.

There are three ways to scan an array – either move the array or move the laser, or both. In general systems that move the array under a fixed laser source are preferred because of their increased light-gathering efficiency.

Another consideration for a confocal laser scanner is the detector system and its associated software for data aquisition. Most commercially available instruments use photomultiplier tubes with amplifiers that allow for high voltage adjustments that can provide up to a 10000-fold sensitivity range, sufficient for accommodating the signal variability and intensity ranges of most array experiments. In addition, the power of the laser can be adjusted. The greater the intensity of the laser beam the more fluorescence; however, this is partially offset by the short time during scanning that each pixel is illuminated. The optimal strength of laser intensity is further dictated by the fluorophore since excessive excitation light can damage the fluorophore and decease the signal-to-noise ratio (photo bleaching). Thus, efforts are under way to develop new fluorophores that can accept more excitation energy, and produce more fluorescence.

Some commercially available scanners are equipped with a charge coupled device (CCD) to detect photons rather than a photomultiplier tube to create an array image. The major drawbacks of CCD systems are the time required for image acquisition and the limited resolution of the image obtained. CCD systems use white light to excite the fluorophores on the array. Since this results in the release of many fewer photons than a focused high-intensity laser source, more time is required to generate an image of acceptable quality. This extended integration time results in higher backgrounds due the generation of dark-current noise. Also, even at a 10 µm resolution the largest CCD chips (approx. 1600 1200 pixels) available at an acceptable price are too small to acquire a single image from a 25 76 mm glass slide array. Thus, laser scanners equipped with photomultiplier tubes are the instruments of choice for most researchers.

All scanners on the market today are controlled through a computer interface to set instrument parameters and retrieve array data. In addition, each manufacturer includes software bundles to filter and process these data, often in proprietary ways that may not fit the experimental designs of every researcher. Thus, given the myriad of ways that individual researchers might wish to process and analyze array data and the rapid advances occurring in this area, it is important to ascertain that the instrument's data aquisition software fits the user's needs and allows retrieval of primary unprocessed data.

Oligonucleotide Microarray

Oligonucleotide arrays are, in effect, miniaturized DNA microarrays and have evolved as an important tool in research into genomes. They are small, high-density arrays, containing tens of thousands of synthetic oligonucleotides. In oligonucleotide microarrays, the probes are short sequences designed to match parts of the sequence of known or predicted open reading frames. Although oligonucleotide probes are often used in "spotted" microarrays, the term "oligonucleotide array" most often refers to a specific technique of manufacturing. Oligonucleotide arrays differ in many details but do share the essential simplicity of the experimental design of DNA microarrays. One type of chip, containing high density arrays of short oligonucleotides, was developed by Affymetrix and makes it possible to array as many as 300,000 oligonucleotides in an area of less than 2 cm^2. These types of chip are known as 'GeneChips'. These are produced by adapting the technique of semiconductor photolithography to synthesise oligonucleotide probe sequences *in situ*. The DNA is synthesised, directly on to the substrate, one base at a time. These strategies generate microarrays of high density using relatively few steps in the synthesis process. The number of probes actually increases exponentially with a linear increase in the number of repetitions of the synthesis cycle. There have been GeneChips made with 96,000 probes per chip. Photolithography has been used to define individual elements of 5–10 µm. This corresponds to an array density of 10^6 probes cm^{-2}.

Fabrication: Oligonucleotide arrays are produced by printing short oligonucleotide sequences designed to represent a single gene or family of gene splice-variants by synthesizing this sequence directly on to the array surface instead of depositing intact sequences. Sequences may be longer (60-mer probes such as the Agilent design) or shorter (25-mer probes produced by Affymetrix) depending on the desired purpose; longer probes are more specific to individual target genes, shorter probes may be spotted in higher density across the array and are cheaper to manufacture. One technique used to produce oligonucleotide arrays include photolithographic synthesis (Agilent and Affymetrix) on a silica substrate where light and light-sensitive masking agents are used to "build" a sequence one nucleotide at a time across the entire array. Each applicable probe is selectively "unmasked" prior to bathing the array in a solution of a single nucleotide, then a masking reaction takes place and the next set of probes are unmasked in preparation for a different nucleotide exposure. After many repetitions, the sequences of every probe become fully constructed. More recently, Maskless Array Synthesis from NimbleGen Systems has combined flexibility with large numbers of probes.

Another approach used is inkjet printing technology

Piezoelectric Printing

- Minute volumes of reagents are delivered to defined locations on the slide similar to 'ink-jet' printing methods.
- The printhead moves across the array, and at each spot electrical stimulation causes the DNA bases, cDNAs or other molecules to be delivered on to the surface via tiny jets.
- This is a non-contact process.

Photolithography

This makes use of semiconductor technologies.

- A mercury lamp is used, and the light activates modified photospecific versions of the four DNA bases.
- Photolithographic masks ensure the DNA synthesis is stimulated in defined

positions—the masks predetermine which of the nucleotides are activated.
- This 'in situ' fabrication technique was developed by Affymetrix, and is used to produce their GeneChips.

Uses of oligonucleotide arrays

- Oligonucleotide arrays have a wide range of possible applications, and in a wide range of species.
- In many cases, they have been employed for the same purposes as DNA microarrays.
- The main areas of application in which oligonucleotide arrays can speed up genomics studies are:
  - Genotyping
  - Gene expression
  - Research into diseases
  - Mapping genomic libraries

cDNA Microarray

cDNA microarray technology is considered one of the most important and powerful tools used to extract and interpret genomic information. The cDNA microarray experiment requires to isolate ribonucleic acid (RNA) from both control (known) and experimental (patient) samples. The reverse transcription process is used to convert the extracted RNAs into cDNAs, which are further labeled with fluorescent probes, usually Cy3 for the control and Cy5 for the experimental channel. After subsequent hybridization and washing procedures, cDNA microarrays are scanned at the ~540 nm (green) for the control and ~630 nm (red) for the experimental channel respectively. The scanning procedure produces two 16-bit monochromatic images, which are further registered into a two-channel, red-green image. Analysis of cDNA microarray data helps in monitoring the expression levels of thousands of genes simultaneously and provides information relevant to cell activity. Therefore, cDNA microarrays have found applications in toxicological research, gene and drug discovery, and disease diagnosis (e.g. cancer, diabetes, and genetic diseases).

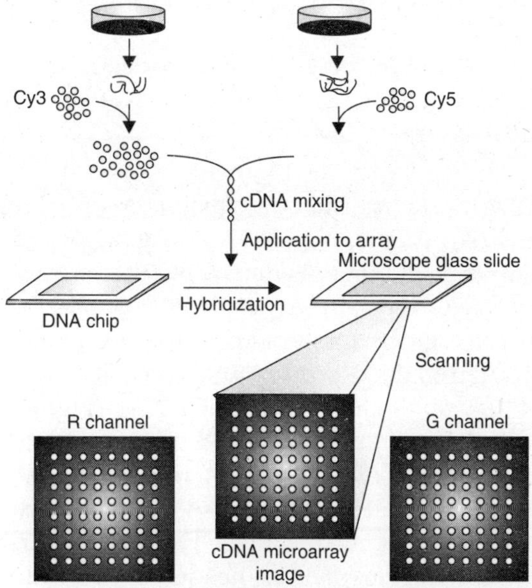

Fig.11.6: cDNA microarray construction

Fabrication: First step of manufacturing a glass cDNA microarray is selecting the material to spot on to the microscope glass surface followed by preparation and purification of DNA sequence representing gene of interest. In the preparation process, PCR can be used to amplify the DNA from the library of interest using a universal primer or gene specific primer and the purity of the DNA fragments representing genes of interest are generally checked by sequencing or agarose gel electrophoresis. This is the important step because all the DNA fragments should be similar concentration and size to achieve similar reaction kinetic for all hybridisation. Next step is spotting DNA solution on to chemically modified glass slide usually with poly-l-lysine or diazotization chemistry and dendrimeric structure. In the case of polylysine, negatively-charged phosphate group in the DNA molecule form an ionic bond with positively charged amine derivative surface. The last step of manufacturing glass DNA microarray is post printing process step involving the drying of the DNA on the slide overnight at room temperature and the UV cross-linking to prevent subsequent binding of DNA and to decrease the background signal upon hybridisation of a labeled target.

The spots of cDNA image constitute the foreground information which is essential for microarray image analysis and gene expression tasks. Red spots determine the presence of RNA from the experimental population of cells, green spots indicate the presence of RNA from the control population of cells, and yellow spots determine that RNAs originate from both experimental and control populations. Arrays of cDNA spots, usually up to 80 000 probes per 2 4 cm^2 area, are commonly referred to as microarrays. The vast amount of data and calculations needed to obtain the relative expression levels of the genes from the fluorescence intensity at each spot necessitates the development of automated data processing solutions.

Tissue Chip

Tissue microarrays (also TMAs) consist of paraffin blocks in which up to 1000 separate tissue cores are assembled in array fashion to allow multiplex histological analysis. Tissue microarrays (TMAs) are used to analyze

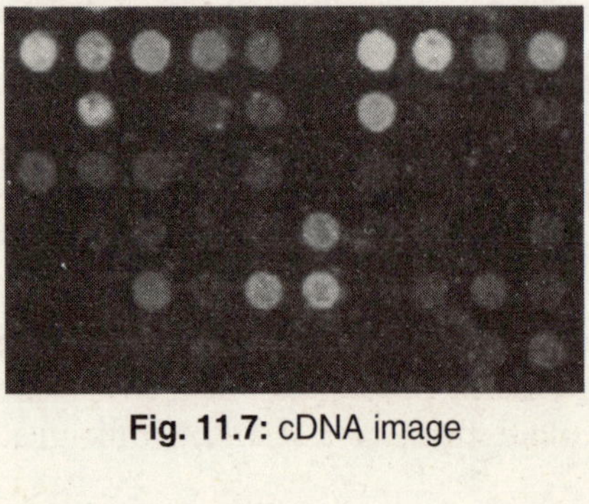

Fig. 11.7: cDNA image

the expression of genes simultaneously in multiple individual tissue samples on one slide. TMAs are composed of small 0.6–3.0 mm cores of tissue from donor tissue paraffin blocks which are arrayed at a high density on a slide. Tissue microarrays have allowed tissue traditional analysis of conventional histological paraffin blocks to become miniaturized and high-throughput.

Histochemical and molecular detection techniques can be used on the TMA slides, and thus allow TMAs to be powerful tools to examine gene expression profiling in disease states across a variety of patients and disease conditions.

Fabrication:

- Select the donor blocks, choosing blocks containing specimen that are approximately equal in thickness.
- Prepare a reference section(3 µm) freshly stained with hematoxylin and eosin (H and E stained slide).
- Prepare a blank paraffin block for use as a recipient for the tissue sample—melt standard embedding paraffin and pour it into an embedding mold, place a plastic block holder on top of the melted paraffin, allow the complex to cool, and then separate the mold from the solidified donor block.
- Design a layout of TMA (tissue microarray) to accommodate the number of donor tissue to be arrayed in the TMA.
- Identify the selected region of tissue using reference section stained with H and E as to guide to ensure that the tissue biopsies are retrieved from the areas previously identified for analysis. Once the desired region is identified, remove the H and E stained slide and retrieve the sample.
- Retrieve the tissue cores using the arraying instrument: Make a hole in the recipient block, acquire a sample of tissue from the

donor block, deposit this sample into its corresponding hole in the recipient block.

- Place the array block on a level surface, in preparation for sectioning-Heat the array block by keeping it in an incubator for 15 min at 37 °C. After heating place glass slide on top of array block and apply pressure to push all tissue cores on the array to the same level. Cool the array block to room temperature.
- Section the array block by cutting 5 µm sections to generate multiple TMAs. Cool the block between -10°C and +10°C. Float the section on to sterile distilled water.
- Bake the TMAs for 30 min at 60°C, remove the excess paraffin and moisture and to promote adherence of the TMA to the microscopic slide and can be stored for 1year in moisture free environment.

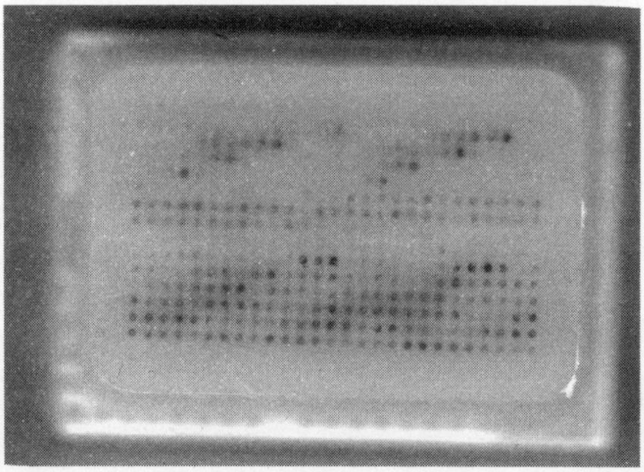

Fig. 11.8: Tissue microarray

Applications of tissue microarrays:

- Immunohistochemistry.
- Staining of : H and E, HC, ISH.
- In situ hybridization.
- Fluorescent in situ hybridization (FISH)
- In situ PCR.
- RNA or DNA expression analysis.
- TUNEL assay for apoptosis.
- Morphological and clinical characterisation of many patient tissues.

Advantages of tissue arrays or tissue chips

- Experimental uniformity.
- Allows amplification (and conservation) of scarce tissue samples.
- Saving the assay volume.
- Reduces the number of slides examined (conventional slides tissue paraffin).
- High-throughput.
- Analysis can be automated and data can be computerized.

Proteinchip

A protein microarray, sometimes referred to as a protein binding microarray, provides a multiplex approach to identify protein-protein interactions, to identify the substrates of protein kinases, to identify transcription factor protein-activation, or to identify the targets of biologically active small molecules. The array is a piece of glass on which different molecules of protein or specific DNA binding sequences (as capture probes for the proteins) have been affixed at separate locations in an ordered manner thus forming a microscopic array. The most common protein microarray is the antibody microarray, where antibodies are spotted onto the protein chip and are used as capture molecules to detect proteins from cell lysate solutions.

Protein microarrays are now becoming very popular due to their possible future applications in the study of nucleic acid–protein, protein–protein, ligand–receptor, drug–protein target, and enzyme–substrate interactions

Applications: Protein microarrays (also biochip, proteinchip) are measurement devices used in biomedical applications to determine the presence and/or amount (referred to as relative quantitation) of proteins in biological samples, e.g. blood. They have the potential to be an important tool for proteomics research. Usually, a multitude of different capture agents, most frequently monoclonal antibodies, are deposited on a chip surface (glass or silicon) in a miniature array. This format is often also referred to as a microarray (a more general term for chip based biological measurement devices).

Types of chips: There are several types of protein chips, the most common being glass slide chips and nano-well arrays.

Production of protein arrays: The production process depends on the type of protein chip.

Protein–protein array: The proteins can be externally synthesised, purified and attached to the array. Alternatively, they can be synthesised *in situ* and directly attached to the array. The proteins can be synthesised through biosynthesis, cell-free DNA expression or chemical synthesis. *In situ* synthesis is possible with the latter two. With cell-free DNA expression, proteins are attached to the support right after their production. Peptides chemically procured by solid phase peptide synthesis are already attached to the support. Selective deprotection is carried out through lithographic methods or by the so-called SPOT-synthesis.

DNA-protein array: Double-stranded DNA (the exact binding sequence of the protein)is attached/ spotted on the array.

Antibody Microarray

An antibody microarray is a specific form of protein microarrays, a collection of capture antibodies are spotted and fixed on a solid surface, such as glass, plastic and silicon chip for the purpose of detecting antigens. Antibody microarray is often used for detecting protein expressions from cell lysates in general research and special biomarkers from serum or urine for diagnostic applications.

Background: The theoretical background for protein microarray-based ligand binding assays was initially developed by Ekins, et. al. in the late 1980s. According to the model, antibody microarrays not only would permit simultaneous screening of an analyte panel, but would also be more sensitive and rapid than conventional screening methods. Interest in screening large protein sets only arose as a result of the achievements in genomics by DNA microarrays and the Human Genome Project.

The first array approaches attempted to miniaturize biochemical and immunobiological assays usually performed in 96-well microtiter plates. 96-well antibody arrays were first created with 144 elements each for "standard enzyme-linked immunosorbent assays" (ELISA). Similar arrays were used to measure prostate-specific antigen (PSA) and cytokines.

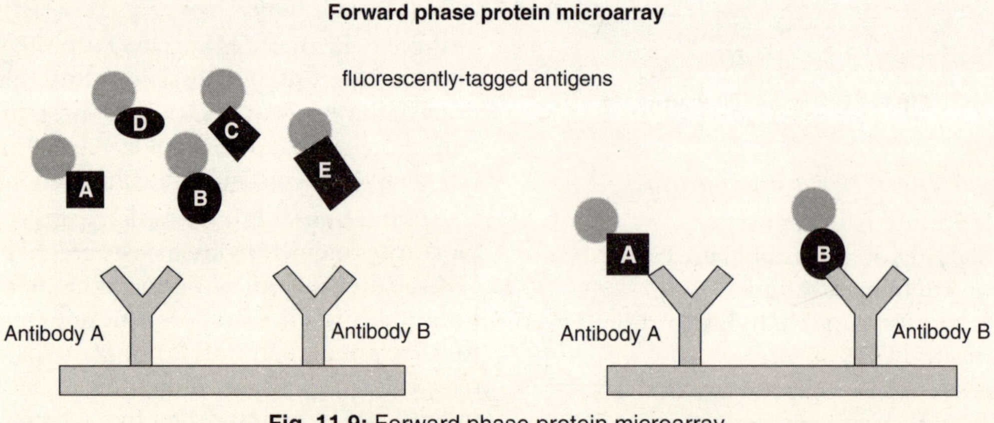

Fig. 11.9: Forward phase protein microarray

Filter membranes were also initially used because of their superior protein binding capacity. They were mostly probed with antibodies using ELISA techniques. A low density array of 48 purified proteins involved in transcription was developed for the investigation of specific interactions of proteins with radiolabeled DNA, RNA, ligands, and other small chemicals. A membrane-based high density array was developed for the purpose of screening a human fetal brain cDNA expression library consisting of 37830 clones. Purified proteins were spotted onto PVDF membranes at a density of 300 samples/cm^2. Other filter-based arrays were constructed but the limitations were the low resolution and considerable background making it difficult to use them in applications with limiting

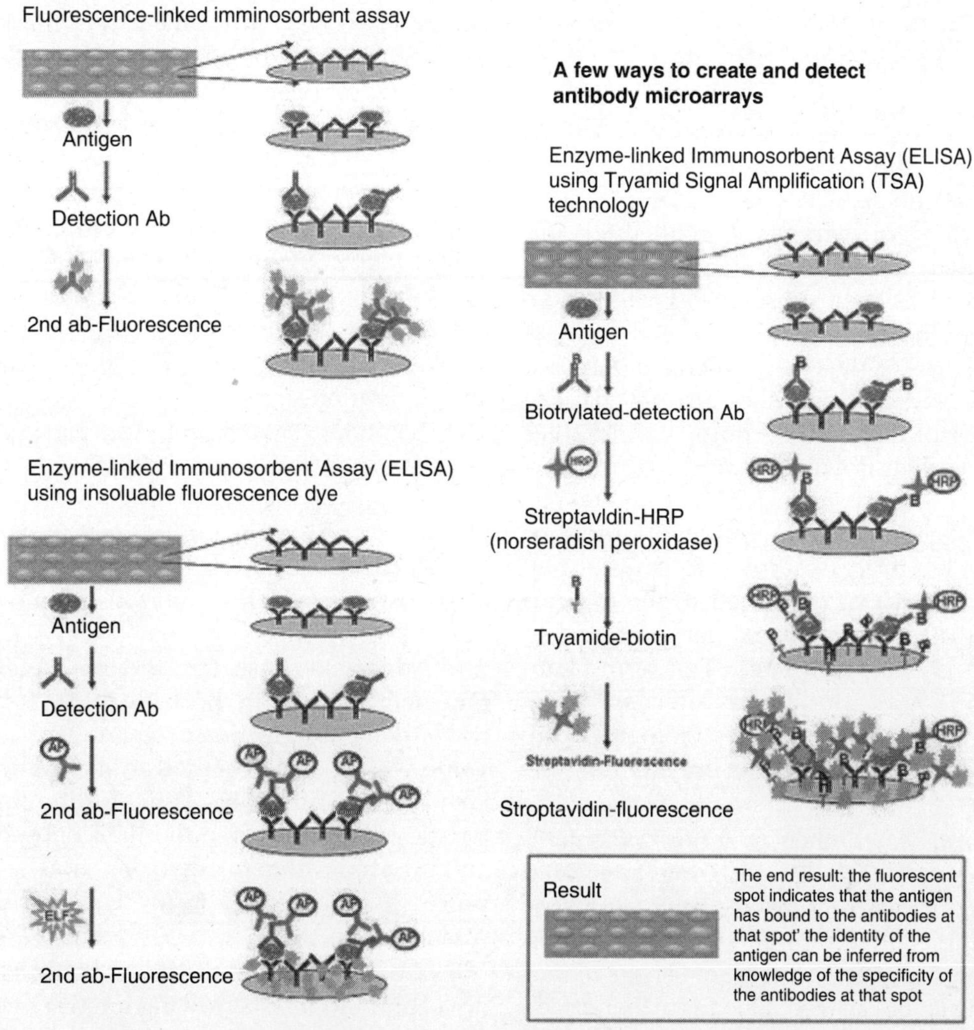

Fig. 11.10: Antibody construction

sample quantities such as protein expression profiling of tumor biopsies.

Protein arrays are comprised of a library of proteins or antibodies immobilized in a 2D addressable grid on a chip. Protein microarray biochips extract and retain targets from liquid media are distinct from microfluidic biochips, which separate and process proteins in a transport medium in situ using microfluidic devices. A typical array may contain 10^3–10^4 spatially distinct elements within a total area of 1 cm^2.

Lab-on-a-chip

A lab-on-a-chip (LOC) is a device that integrates one or several laboratory functions on a single chip of only millimeters to a few square centimeters in size. LOCs deal with the handling of extremtely small fluid volumes down to less than pico liters. Lab-on-a-chip devices are a subset of MEMS devices and often indicated by "micro total analysis systems" (µTAS) as well. Microfluidics is a broader term that describes also mechanical flow control devices like pumps and valves or sensors like flowmeters and viscometers. However, strictly regarded "Lab-on-a-chip" indicates generally the scaling of single or multiple lab processes down to chip-format, whereas "µTAS" is dedicated to the integration of the total sequence of lab processes to perform chemical analysis. The term "Lab-on-a-chip" was introduced later on when it turned out that µTAS technologies were more widely applicable than only for analysis purposes.

History: After the invention of microtechnology (~1954) for realizing integrated semiconductor structures for microelectronic chips, these lithography-based technologies were soon applied in pressure sensor manufacturing (1966) as well. Due to further development of these usually CMOS-compatibility limited processes, a tool box became available to create micrometre or sub-micrometre sized mechanical structures in silicon wafers as well: the micro-electro-mechanical-systems (MEMS) era (also indicated with micro system technology - MST) had started.

Next to pressure sensors, airbag sensors and other mechanically movable structures, fluid handling devices were developed. Examples are: channels (capillary connections), mixers, valves, pumps and dosing devices. The first LOC analysis system was a gas chromatograph, developed in 1975 by S.C. Terry – Stanford University. However, only at the end of the 1980s, and beginning of the 1990s, the LOC research started to seriously grow as a few research groups in Europe developed micropumps, flowsensors and the concepts for integrated fluid treatments for analysis systems.

These µTAS concepts demonstrated that integration of pre-treatment steps, usually done at lab-scale, could extend the simple sensor functionality towards a complete laboratory analysis, including, e.g. additional cleaning and separation steps.

A big boost in research and commercial interest came in the mid 1990's, when µTAS technologies turned out to provide interesting tooling for genomics applications, like capillary electrophoresis and DNA microarrays. A big boost in research support also came from the military, especially from DARPA (Defense Advanced Research Projects Agency), for their interest in portable bio/chemical warfare agent detection systems. The added value was not only limited to integration of lab processes for analysis but also the characteristic possibilities of individual components and the application to other, non-analysis, lab processes. Hence the term "Lab-on-a-chip" was introduced.

Although the application of LOCs is still novel and modest, a growing interest of companies and applied research groups is observed in different fields such as analysis (e.g.

chemical analysis, environmental monitoring, medical diagnostics and cellomics) but also in synthetic chemistry (e.g. rapid screening and microreactors for pharmaceutics). Besides further application developments, research in LOC systems is expected to extend towards downscaling of fluid handling structures as well, by using nanotechnology. Submicrometre and nano-sized channels, DNA labyrinths, single cell detection an analysis and nano-sensors might become feasible that allow new ways of interaction with biological species and large molecules.

Chip materials and fabrication technologies: The basis for most LOC fabrication processes is photolithography. Initially most processes were in silicon, as these well-developed technologies were directly derived from semiconductor fabrication. Because of demands for, e.g. specific optical characteristics, bio- or chemical compatibility, lower production costs and faster prototyping, new processes have been developed such as glass, ceramics and metal etching, deposition and bonding, PDMS processing (e.g. soft lithography), thick-film- and stereolithography as well as fast replication methods via electroplating, injection molding and embossing.

Furthermore the LOC field more and more exceeds the borders between lithography-based microsystem technology, nano-technology and precision engineering

Advantages of LOCs: LOCs may provide advantages, which are specific to their application. Typical advantages are:

- Low fluid volumes consumption (less waste, lower reagents costs and less required sample volumes for diagnostics)
- Faster analysis and response times due to short diffusion distances, fast heating, high surface to volume ratios, small heat capacities.
- Better process control because of a faster response of the system (e.g. thermal control for exothermic chemical reactions)
- Compactness of the systems due to integration of much functionality and small volumes
- Massive parallelization due to compactness, which allows high-throughput analysis
- Lower fabrication costs, allowing cost-effective disposable chips, fabricated in mass production
- Safer platform for chemical, radioactive or biological studies because of integration of functionality, smaller fluid volumes and stored energies

Disadvantages of LOCs:

- Novel technology and therefore not yet fully developed
- Physical and chemical effects—like capillary forces, surface roughness, chemical interactions of construction materials on reaction processes become more dominant on small-scale. This can sometimes make processes in LOCs more complex than in conventional lab equipment
- Detection principles may not always scale down in a positive way, leading to low signal-to-noise ratios
- Although the absolute geometric accuracies and precision in microfabrication are high, they are often rather poor in a relative way, compared to precision engineering for instance.

Examples of LOC applications:

- Real-time PCR; detect bacteria, viruses and cancers.
- Biochemical assays
- Immunoassay; detect bacteria, viruses and cancers based on antigen–antibody reactions.
- Dielectrophoresis detecting cancer cells and bacteria.
- Blood sample preparation; can crack cells to extract DNA.

- Cellular lab-on-a-chip for single-cell analysis.
- Ion channel screening

LOCs and global health: Lab-on-a-chip technology may soon become an important part of efforts to improve global health, particularly through the development of point-of-care testing devices. In countries with few healthcare resources, infectious diseases that would be treatable in a developed nation are often deadly. In some cases, poor healthcare clinics have the drugs to treat a certain illness but lack the diagnostic tools to identify patients who should receive the drugs. Many researchers believe that LOC technology may be the key to powerful new diagnostic instruments. The goal of these researchers is to create microfluidic chips that will allow healthcare providers in poorly equipped clinics to perform diagnostic tests such as immunoassays and nucleic acid assays with no laboratory support.

Global challenges: For the chips to be used in areas with limited resources, many challenges must be overcome. In developed nations, the most highly valued traits for diagnostic tools include speed, sensitivity, and specificity; but in countries where the healthcare infrastructure is less well developed, attributes such ease of use and shelf life must also be considered.

The reagents that come with the chip, for example, must be designed so that they remain effective for months even if the chip is not kept in a climate-controlled environment. Chip designers must also keep cost, scalability, and recyclability in mind as they choose what materials and fabrication techniques to use.

Examples of global LOC application: One active area of LOC research involves ways to diagnose and manage HIV infections. Around 40 million people are infected with HIV in the world today, yet only 1.3 million of these people receive anti-retroviral treatment. Around 90% of people with HIV have never been tested for the disease. Measuring the number of CD4+ T lymphocytes in a person's blood is an accurate way to determine if a person has HIV and to track the progress of an HIV infection. At the moment, flow cytometry is the gold standard for obtaining CD4 counts, but flow cytometry is a complicated technique that is not available in most developing areas because it requires trained technicians and expensive equipment.

MPSS ('massively parallel signature sequencing') Technology

MPSS technology (LYNX proprietary technology) addresses the need to generate sequence information from millions of DNA fragments. At this extremely large scale, the technology approach eliminates the need for individual sequencing reactions and the physical separation of DNA fragments required by conventional sequencing methods. Besides, it enables the simultaneous identification of nearly all the DNA molecules in a sample. Since, the technology is built on the Megaclone technology (LYNX proprietary technology), to better understand MPSS, it is beneficial to elaborate on this technique.

Megaclone Technology

All the functions in the cell are governed by the genes made up of DNA carried in the chromosomes that are located in the cell nucleus. Understanding the operation of these genes requires methods for generating many tens of thousands of DNS fragments in quantities and formats amenable to biochemical analysis. The effort required to analyze DNA fragments involved in genetic processes is greatly reduced by working in parallel with large numbers of very small samples. Megaclone makes use of substrates called microbeads of sizes less than the diameter of human hair. These substrates are suitable for parallel micro-analysis such as MPSS. The following illustration shows how a microbead looks like.

It is shown that the bead is attached with unique 32 bp tags (in orange) which are hybridized with complementary tags attached to the cDNAs (reverse transcript from mRNA in the cells).

As shown in Fig. 11.11 is the final product of megaclone. The mRNAs obtained from the cells are firstly reverse transcribed to generate corresponding cDNAs. These cDNAs are then hybridized with 32 bp unique tags. The cDNA-tag combinations are simultaneously amplified by polymerase chain reaction (PCR) to generate large number of each combination. As for the microbeads, they are all attached with many copies of unique anti-tag sequence complementary to the tags on the cDNA. The two components (cDNA-tags and microbeads) are added together and hybridization is allowed to take place. All reactions are simultaneously carried out in a small test tube. The following illustration shows the flow of the process of generating megaclone:

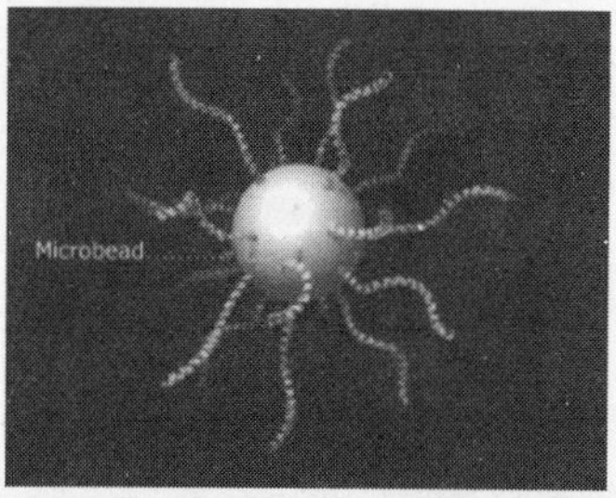

Fig. 11.11: Microbeads used in megaclone

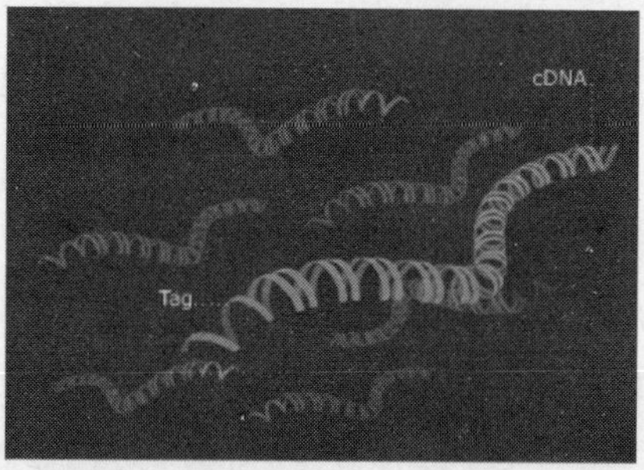

a. cDNA tagged with 32 bp unique tags and amplified with PCR

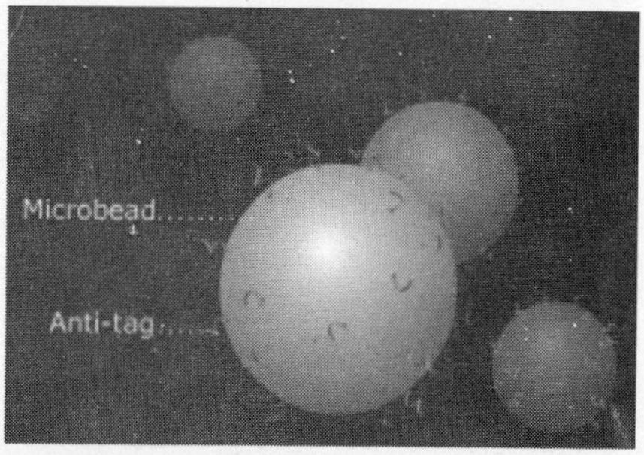

b. Microbeads tagged with many anti-tags complimentary with the tags on the cDNAs

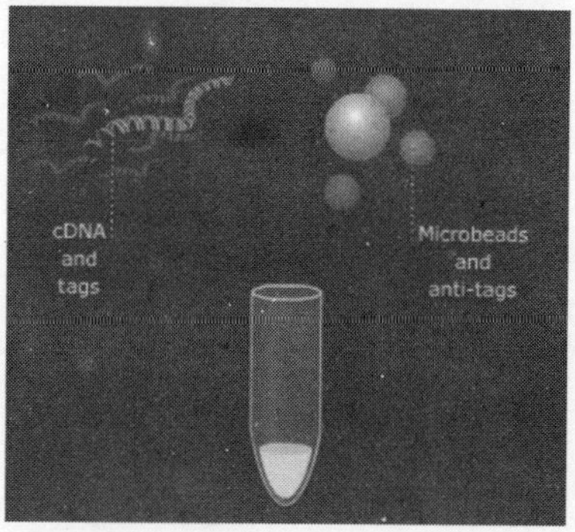

c. Both components from a) and b) are mixed together to allow hybridization of the anti-tags on the beads and the tags on the cDNAs

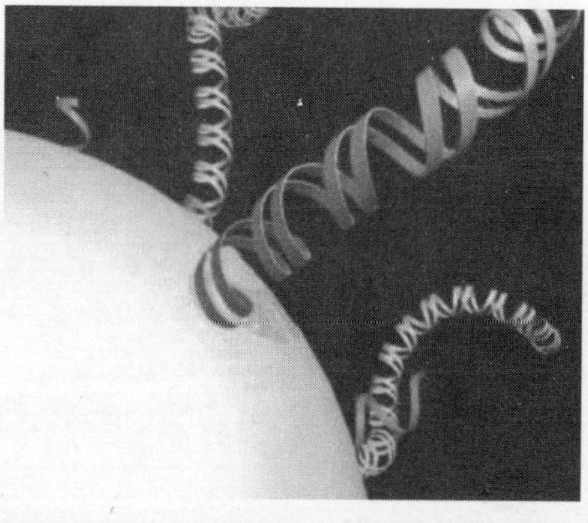

d. The final product with the tags on the cDNA hybridizied to the anti-tags of the microbead

Figs 11.12 (a to d): Simplified flow of megaclone technology

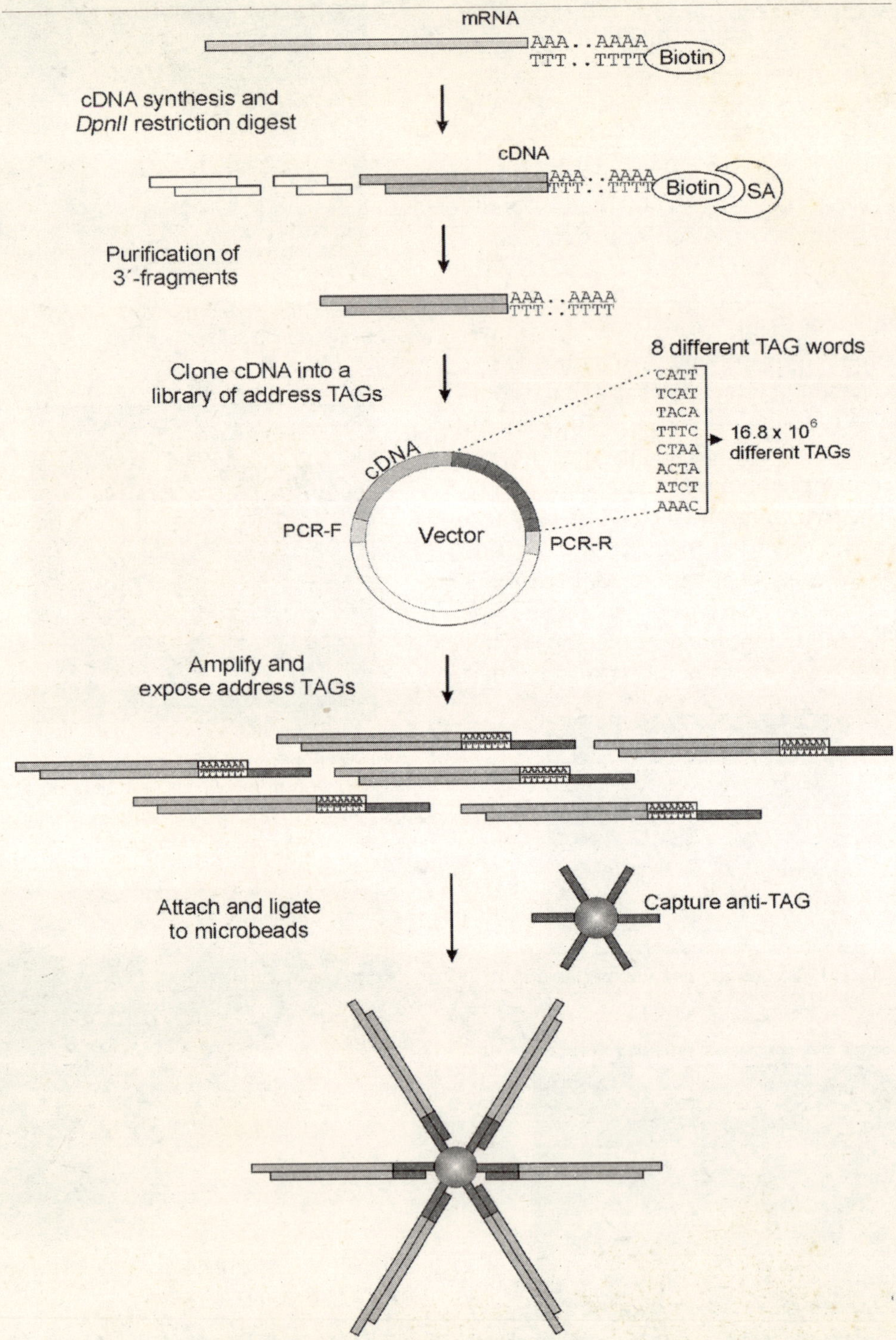

Fig. 11.13: Summary of megaclone technology

Poly (A) mRNA molecules are converted into double stranded cDNA molecules, which are digested with Dpn II and cloned into a specially designed plasmid vector containing a 32 base pair (bp) oligonucleotide tag. There are 16.8 10^6 million different 32-base sequences available in the reference tag library, and each cDNA clone contains a different sequence. A library of cDNA inserts, along with their adjacent 32 bp oligonucleotide tags, are polymerase chain reaction (PCR) amplified, and the resulting linear molecules are partially treated with an exonuclease to make the 32- base tag single stranded. The 32-base tags at the end of each of the cDNA molecules are hybridized to 32-base.

As mentioned, the megaclone is the front phase of the MPSS technology. Once the solution of the hybridized beads and cDNA is obtained, we will perform the sequencing process which is very much different from the conventional technique. The sequencing reaction takes place in a glass chamber referred to as a flow cell (Fig. 11.15). The flow cell is fabricated by micro-machining a glass plate to form a grooved chamber for immo-

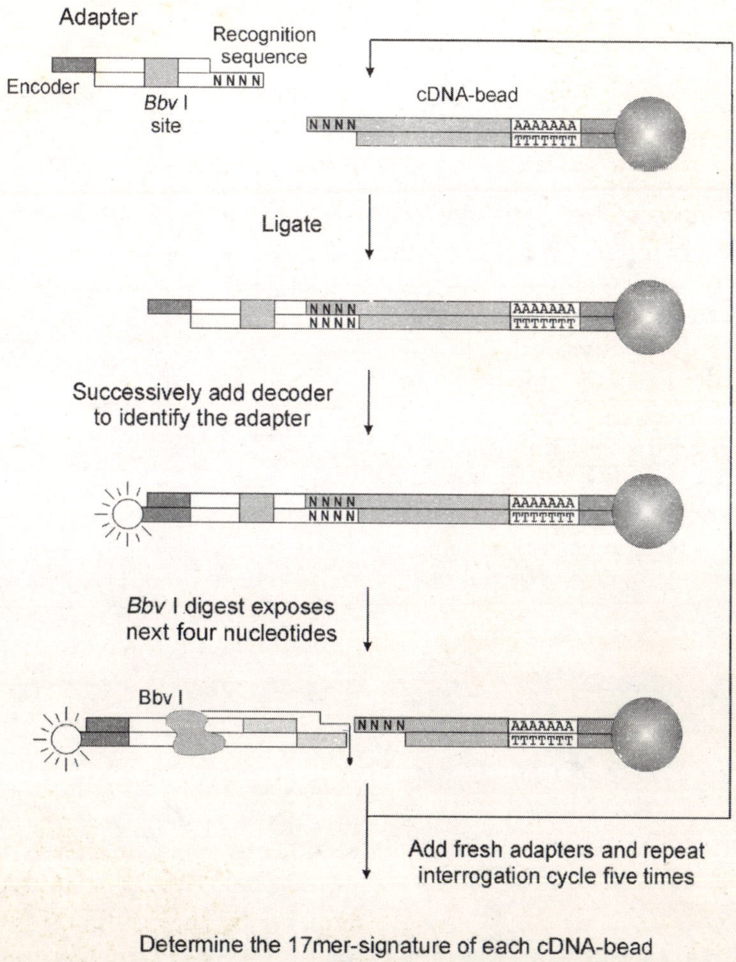

Fig. 11.14: Summary of the sequencing reactions for MPSS

11

bilizing as many as two million megaclone beads in a planar array. As the microbeads flow in through the inlet, they travel along the grooves of the flow cell and pack against a vertical constriction, or dam adjacent to the outlet. The dam stops the beads, but allows enough head space for reagents to flow through the chamber. The microbeads are fixed and arrayed in a monolayer, allowing the software to track each megaclone bead individually as the sequencing proceeds. The sequencing cycle begins by flowing reagents in through the inlet, passing the reagents over each bead, and exiting through the outlet. The sequencing process takes advantage of a special property of a type IIs restriction endonuclease which its cleavage site is separated from its recognition site by a characteristic number of nucleotides.

Encoded adaptors are ligated to the ends of the cDNA molecules attached to the microbeads. Sixteen different fluorescent-labeled decoder probes are then sequentially hybridized to the encoded adaptor ends in order to deduce the first four nucleotides to the adaptor ends in order to deduce the first four nucleotides at the end of each molecule. The encoded adaptor from the first round is then removed digestion with BbvI, which exposes the next four nucleotides as a four-base single stranded overhang. The process is repeated several times in order to generate a total of 17 bases of signature sequence.

Fig. 11.15: Illustration of the flow cell where all the reactions take place

Dyes used for Labeling

Cy3 and Cy5: Cy3 and Cy5 are reactive water-soluble fluorescent dyes of the cyanine dye family. Cy3 dyes are green (~550 nm excitation, ~570 nm emission), while Cy5 is fluorescent in the red region (~650/670 nm). They are usually synthesized with reactive groups on either one or both of the nitrogen side chains so that they can be chemically linked to either nucleic acids or protein molecules. Labeling is done for visualization and quantification purposes. They are used in a wide variety of biological applications including comparative genomic hybridization and in gene chips, which are used in transcriptomics. They are also used to label proteins and nucleic acid for various studies including proteomics and RNA localization.

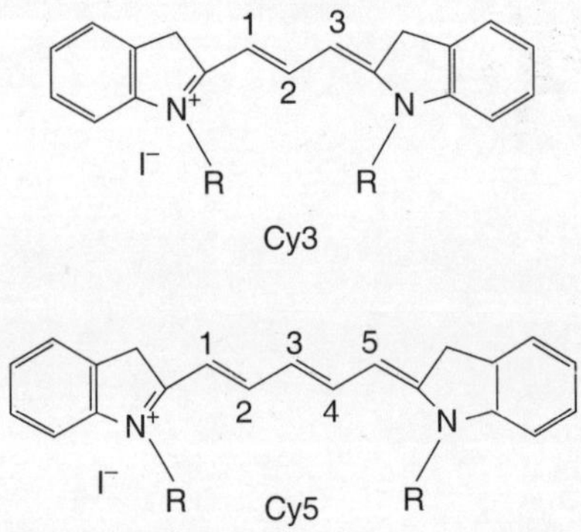

Fig. 11.16: Structures of Cy3 and Cy5

Spectral characteristics: Cy3 is excited maximally at 550 nm and emits maximally at 570 nm, in the green part of the spectrum; quantum yield in PBS buffer is 0.04; FW=766.

Cy5 is excited maximally at 649 nm and emits maximally at 670 nm, in the far red part of the spectrum; quantum yield is 0.28. FW=792.

The scanners actually use different laser emission wavelengths (typically 532 nm and 635 nm) and filter wavelengths (550–600 nm and 655–695 nm) to avoid background contamination. They are thus able to easily distinguish between two samples when one sample has been labeled with Cy3 and the other labeled with Cy5. They are also able to quantitate the amount of labeling in either sample.

Cy5 ozone susceptibility: In 2003, researchers at Inpharmatics and Agilent reported in analytical chemistry that microarrays which used Cy5 were susceptible to intermittent data quality decrease caused by environmental ozone. Exposures to ozone levels above 5-10 ppb for 10–30 seconds were reported to decrease the reproducibility of Cy5 microarrays. Much higher levels of ozone (>100 ppb) were required to observe an effect in Cy3.

Alexa fluor: It is a reactive dye used in microarray marketed by Invitrogen (Table 11.2).

Enhanced fluorescence-based assays

DNA dendrimer technology: One pathway for achieving higher detection sensitivity is to increase the number of labels associated with each cDNA or target nucleic acid bound to a microarray. A number of research groups have explored using branched- or dendrimer-based nucleic acid structures to increase the label density per nucleic acid target. For microarray labelling, Genisphere has developed fluorophore labelled nucleic acid dendrimers which are referred to as 3DNA probes. The 3DNA probes are prepared by hybridizing and cross-linking complementary oligonucleotide building blocks to form a 'core' dendritic structure. Cy3 or Cy5 labeled oligonucleotides are subsequently hybridized and cross-linked to the core dendrimers so that each 3DNA probe contains on average ~ 250 fluorophore dyes. For gene expression monitoring, cDNAs are coded with a single universal sequence tag during transcription which is recognized by a 3DNA probe containing the universal code complement in a second labeling step.The 3DNA probe yields 250 fluorophores/cDNA target compared to ~12 fluorophores/cDNA target through direct incorporation of fluorophore labeled dNTPs. This corresponds to a ~20-fold increase in the number of fluorophores/cDNA probe over direct enzymatic incorporation. For comparison of experimental detection limits, 1–50 µg of total RNA was transcribed and labelled with both methods. After hybridization to cDNA arrays, the average specific signal for each gene was measured. The specific signal obtained with the 3DNA dendrimer probe using 2.5 µg of total RNA was equivalent to the specific signal obtained with direct Cy3 incorporation using 40 µg of total RNA. This amounts to a ~16-fold increase in detection sensitivity which correlates well with the number of fluorophores bound per probe. In addition, it was noted that the 3DNA probe signal was stable to repetitive scanning whereas the Cy3 labeled cDNA signal significantly degraded over time, demonstrating that this labeling technology is more robust. In summary, the

Table 11.2: Alexa fluor

Catalog#	*Fluorescent dye*	λ_{max} *(nm)**	*Em (nm)†*	ε_{dye} $(cm^{-1}M^{-1})$‡	CF_{260}§	*Spectrally similar dyes*
A32750	Alexa fluor	495	519	71.000	0.3	Cy2, fluorescein (FAM), fluorescein isothiocyante (FITC)
A32756, A32755	Alexa fluor® 555	555	565	150.000	0.04	Cy3, tetramethylrhodamine (TRITC, TAMRA)
A32751	Alexa fluor® 594	588	588	80,400	0.43	Texas Red
A32757 A32755	Alexa fluor® 647	650	650	239,000	0.00	Cy5

* Absorbance maximaum for the dye, † Emission maximum for the dye, ‡ Extinction coefficient for the dye, § Correction factor = A_{260} for the free dye/A_{max} for the free dye

use of significantly less RNA starting material, the higher stability of the label compared with conventional fluorophores, and the ready integration of the labels with existing fluorescence microarray scanners make the 3DNA dendrimer labelling technology attractive for gene expression analysis (Fig. 11.17).

Semiconductor quantum dots: Semiconductor quantum dots (QDs) have emerged as a new class of fluorophore labels. These new labels comprise nanometer sized particles of group II–VI or III–V atoms from the periodic table of elements such as CdSe or InAs that are smaller than the exciton Bohr radius (typically 1–10 nm in size). As a result, the QDs exhibit quantum confinement effects resulting in optical properties that are significantly different than the corresponding bulk material or the atoms that comprise the particle. The fluorescence emission of the particles may be tuned from blue to the near infrared by controlling particle size and chemical composition, which alters the band gap of the particles. Narrow fluorescence emission bandwidths (25–30 nm FWHM for CdSe QDs) have been observed which makes it possible to generate many spectrally unique QDs for multiplexing applications. Importantly, the particles exhibit broad light absorption that occurs from the ultraviolet through the lowest energy band gap, and as a result, multicolored QDs may be excited by a single UV light source. QDs have exhibited quantum yields of 40–50%, which are slightly lower than the quantum yields for commercial organic fluorophores.

This is compensated by the high molar extinction coefficients of such particles at 10^5–10^6 M^1 cm^1, which is 10–100 times larger than that for typical organic fluorophores. The QDs are also highly stable against photobleaching.

These fluorescence properties are suitable for many biological applications such as cellular imaging and fluorescent in situ hybridization (FISH) where stable fluorophores are desired, and microarray labeling where multiple colored probes are possible using a single excitation source.

Phosphor reporters: Autofluorescence background on microarray substrates negatively impacts the sensitivity of fluorophore labeling. A novel approach devised to eliminate autofluorescence utilizes up-converting phosphor labels that absorb two photons of lower frequency light in the infrared region and emit a single photon at a higher frequency in the visible region. Up-converting phosphor (UP) labels typically comprise submicron-sized yttriumYoxysulfide par-

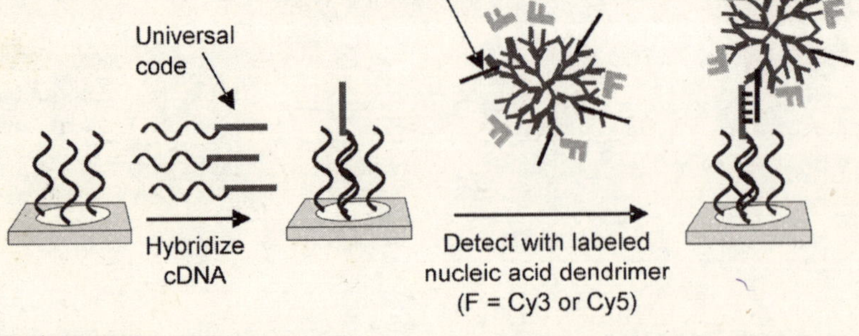

Fig. 11.17: Schematic illustration of cDNA detection on microarrays using fluorophore labeled dendrimers. Note that dendrimers typically contain ~250 fluorophores

ticles (0.2–0.4 μm diameter) that are doped with lanthanide ions such as ytterbium and erbium for excitation and emission. Phosphorescence from the lanthanide ions persists for > 10^{8} seconds after excitation while organic fluorophores emit light for < 10^{9} seconds after illumination; thus, phosphorescence is distinguishable from autofluorescence using time resolved fluorescence spectroscopy. Up-converting phosphor labels with different emission colors are generated by using the same absorber ion with different lanthanide ions for emission. For example, the ytterbium/erbium excitation/emission pair emits green light while the ytterbium/thulium pair emits blue light. UP materials also are characterized by narrow emission bandwidths (25–50 nm), which has enabled the development of over six spectrally unique emission colours. Additionally, infrared excitation is advantageous for microarrays since other assay components (e.g. substrates) do not absorb infrared light resulting in lower overall background. A detection limit of ~12 UP particles in a 30–40 mm^2 well was achieved using infrared laser excitation and detection with a photomultiplier tube. Therefore, the theoretical detection limit of a few UP labels per microarray spot is orders of magnitude better than molecular Cy3 fluorophores with detection limits of ~ 1–5 fluors/μm^2.

Surface-enhanced Raman Spectroscopy (SERS)

The Raman effect occurs when light impinges upon a molecule and interacts with the electron cloud and the bonds of that molecule. The incident photon excites the molecule into a virtual state. For the spontaneous Raman effect, the molecule will be excited from the ground state to a virtual energy state, and relax into a vibrational excited state. Two series of lines exist around this central vibrational transition. They correspond to the complimentary rotational transition. Anti-Stokes lines correspond to rotational relaxation whereas Stokes lines correspond to rotational excitation.

A change in the molecular polarization potential — or amount of deformation of the electron cloud — with respect to the vibrational coordinate is required for the molecule to exhibit the Raman effect. The amount of the polarizability change will determine the Raman scattering intensity, whereas the Raman shift is equal to the vibrational level that is involved.

Surface-enhanced Raman spectroscopy (SERS) is a Raman spectroscopic (RS) technique that provides greatly enhanced Raman signal from Raman-active analyte molecules that have been adsorbed on to certain specially prepared metal surfaces. Increases in the intensity of Raman signal have been regularly observed on the order of 10^4–10^6, and can be as high as 10^8 and 10^{14} for some systems.

The importance of SERS is that it is both surface selective and highly sensitive where as RS is neither. RS is ineffective for surface studies because the photons of the incident laser light simply propagate through the bulk and the signal from the bulk overwhelms any Raman signal from the analytes at the surface. SERS selectivity of surface signal results from the presence of surface enhancement (SE) mechanisms only at the surface. Thus, the surface signal overwhelms the bulk signal, making bulk subtraction unnecessary.

Raman spectroscopy is closely related to infrared absorption spectroscopy.

- Provides fingerprint spectra (molecular identity, sample composition)
- Complementary to IR
- Sensitive to molecular conformation, environment
- Laser-based (easy to focus beam); diffraction-limited spatial resolution

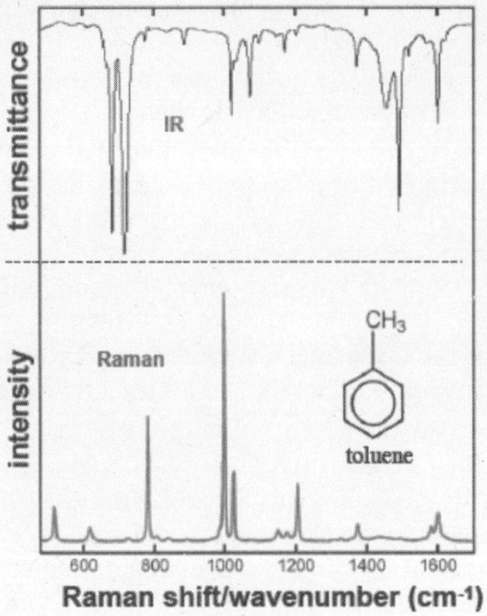

Fig. 11.18: Raman spectra

- Useful for biological applications
  - Non-destructive
  - Analytes do not photobleach
  - Provides contrast without specific labelling.

Different molecules are characterized by their own, unique Raman signature.

Applications of SERS

1. To identify presence of a carbonate or phosphate in proteins.
2. To differentiate sulfhydryl group (-SH) and disulfide group(-S-S) often involved in converting a protein from inactive to active form (Ex—insulin)
3. Binding studies
   - Protein- protein binding
   - Protein-nucleic acid interaction
   - Hormone-receptor binding
   - Protein drug interactions
4. Detection of biomolecules.
5. Detection platforms for biological toxins.
6. Cancer gene detection.
7. Pathogen detection.

Applications of Microarray Technology

Biological Warfare

Speed is the essence in successfully containing a biological warfare attack. Quickly identifying the agent and how to treat those who have been exposed are keys to controlling an outbreak and minimizing its destructiveness. A handled device containing a lab on chip may just be the answer. The result of breakthroughs in biology, chemistry and micro-manufacturing, the instrument can immediately alert investigators to even the slightest hint of anthrax or smallpox in the air. At Massachusetts Institute Of Technology biologist Jianzhu Chen and his colleagues hope to deliver a prototype biosensor in less than 18 months. The work is part of defense advanced Research Projects Agency's four year, $ 24-million tissue-based biosensor program, which funds research by about a dozen universities and private firms. Mouse B cells power the device. Part of immune system, B cells express antibodies on their surfaces that bind to particular infectious particles. For example, most humans harbor B cells for pathogens that cause colds, polio, tetanus, and other diseases. When a B cell binds to the intruder that is built to recognize, a biochemical cascade occurs in the cell, triggering the body's immune response to rally to the defense. B cells have been genetically engineered to respond to particular biological warfare agents. To know that the B cells have actually gone into action, the researchers plug in another gene – from jellyfish called Aequorea. This gene enables the gelly fish to glow with bioluminescent protein aequorin. The aequorin instantly emits light when triggered by calcium ions – a substance that is produced when the bioagent-induced cascade occurs in the B cell. The entire process, from detection to bioluminescence, takes less than a second, beating any human handiwork to date. Other methods have matched either the speed or the

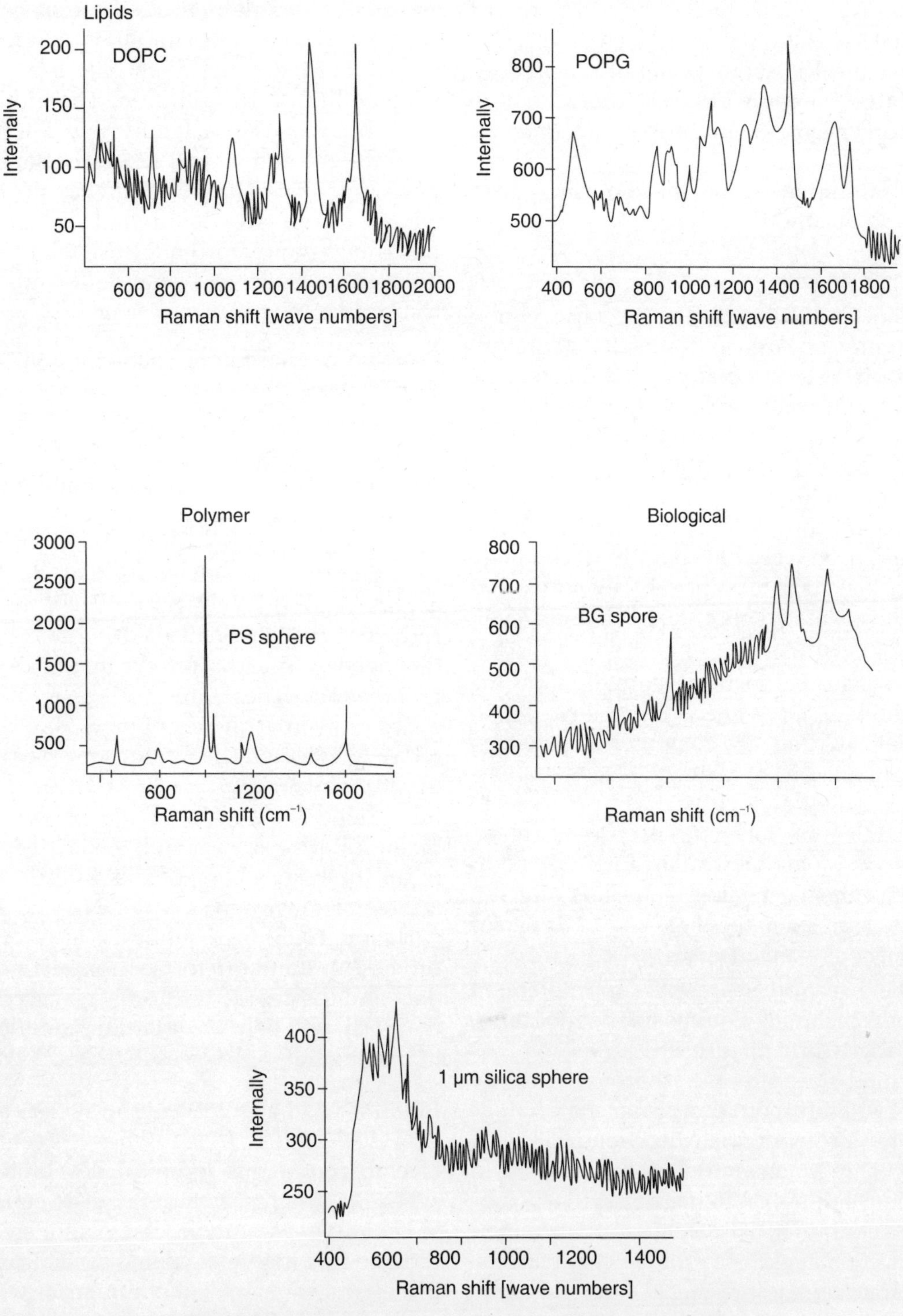

Fig. 11.19: Unique Raman signature for different molecules

sensitivity of B cells, but not both. The record for analysis using polymerase chain reaction of a bioagent is about 12 minutes, based on a pristine sample containing more than 20 microorganisms. Immunoassay techniques, which also use an antibody-capture methodology, are approaching the requisite speed but lack sensitivity.

Forensic Assays

The biochip created by the biochip technology center at Argone National Laboratory and their sister laboratory, The Englehardt Institute of Molecular Biology (Moscow, Russia) is being developed for use in forensic DNA typing. The biochip called as MAGIC Chip (Microarrays of Gel – Immobilized Compounds on a chip). This is a collaborative effort between the Biochip Technology Center and Argone National Laboratory and the Illinioís State Police Research and Development Laboratory.

This Lab on Chip may eventually allow the portability of DNA analysis possibly even at the site of Crime, like computer chips which perform millions of mathematical operations per second. MAGIC Chip has been used to test strains of tuberculosis, breast cancer genes and biological warfare agents. In addition biochips containing thousands of genes is being utilized in the discovery of drugs for treatment of many diseases.

The goal of this project is to implement this new technology in the analysis of mitochondrial DNA for use in human remains cases and forensic cases that contain only hairs (without root or follicular material) as the physical evidence. Mitochondrial DNA sequencing is currently being validated in Illinois State Police Research and Development Laboratory. This technology is of interest in forensic science for several reasons. First, mt DNA is present in large amounts in all the cells. This means that only a small amount of sample is needed for successful analysis. Second, human mt DNA is maternally inherited, only the mother's mt DNA is passed on to the children. Therefore, all individuals within a family that are related through mother will share the same mt DNA type. This is important when the DNA analyst is trying to identify human remains. If a known sample from a close relative can be used as a reference sample for mt DNA testing.

Promise of Microarray Technology in Treating Disease

Now that you understand the concept behind array technology, picture this: a hand-held instrument that a physician could use to quickly diagnose cancer or other diseases during a routine office visit. What if that same instrument could also facilitate a personalized treatment regimen, exactly right for you? Personalized drugs, molecular diagnostics, integration of diagnosis and therapeutics. These are the long-term promises of microarray technology. May be not today or even tomorrow, but someday these may turn into reality. For the first time, arrays offer hope for obtaining global views of biological processes, simultaneous readouts of all the body's components by providing a systematic way to survey DNA and RNA variation. NCBI, by continuing its efforts to provide a standard format for microarray data and to provide free, universal access to that data, will help the scientific community in making those promises realities.

Application of Microarray in Development of Personalized Medicine

Due to continuous technical developments and new insights into the high complexity of many diseases, molecular pathology is a rapidly growing field gaining center stage in the clinical management of tumors as well as in the pharmaceutical development of new anti-cancer drugs. The application of novel compounds in clinical trials has revealed promising results; however, the current diagnostic procedures available for determining which patients will primarily benefit from rational tumor therapy are insufficient.

To read a patient's tissue as "deeply" as possible, in the future, gaining information on the morphology and on genetic, proteomics, and epigenetic alterations will be the upcoming task of surgical pathologists experienced in molecular diagnostics to provide the clinicians with information relevant for an individualized medicine.

Among the different high-throughput technologies, DNA microarrays are now the first array approach close to enter routine diagnostics. Technically advanced and well-established microarray platforms can nowadays be evaluated by distinct bioinformatics tools capable of identifying both novel genes associated with disease development and clusters of genes predicting clinical outcome of an individual tumor.

The automatic, highly parallel analysis of proteins and complex proteins lysates for early detection of cancers such as breast, prostate and ovary as proteomic patterns in the serum also appears at the horizon. In addition, an improved analysis of tumor samples via antibody or reverse-phase protein arrays is likely to provide the pathologist in the future with information about activated oncogenic signaling pathways and other cell functions, such as drug response or the potential to metastasize.

While expression microarrays and proteomic analysis rely on relatively unstable material incompatible with paraffin-embedded tissue samples, an investigation of DNA methylation using specialized high-throughput platforms has revealed the potential of being used in future diagnostics. Each of these approaches on its own might not suffice to extract all information required for an efficient individualized diagnostics. Therefore, a "multiplex approach" combining the different biological levels DNA, RNA, and protein, may be necessary to functionally classify malignant tumors. This appears to become a major challenge for diagnostic pathologists.

Application of Microarray in Pharmacogenomics

The DNA microarrays have proven to be a state of the art technique for high throughput comprehensive analysis of thousand of genes in parallel. The application of a DNA microarray to compare normal and pathological cells, tissues or organs may allow, along with classical positional cloning techniques, to speed up the discovery of genes and gene pathways implicated in several diseases.

This in turn will result in further application of the DNA microarray technique in the field of pharmacogenomics in order to characterize and validate new therapeutic targets, their mechanism of action, metabolic pathways and unwanted secondary effects. However, the lack of standardized criteria for the analysis and interpretation of the huge amount of data generated by DNA microarrays may hamper its utilization on a routine basis in the human health domain.

Microarray in Cancer

Cancer is one of the leading causes of death in the modern world. The term cancer is used when cells lose their ability to regulate growth, become immortal, and in most cases migrate from their primary location into a new unrelated site. Among cancers, prostate cancer is the leading cause of cancer-related death among men in America, next only to lung cancer. Prostate cancer is clinically subdivided into various classes based on its progression. These include benign, localized, and metastatic (spread into other tissues) prostate cancer. Of these, surgical and radiation therapies exist for clinically localized prostate cancer. However, once the cancer becomes metastatic, it becomes incurable.

With the advent of microarray technology, it has become possible to study global gene expression profiles. However, a comparative global protein expression analysis for cancer tissues has not been accomplished to date. Antibody microarray can be used to look for changes in levels of proteins during different stages of cancer. This slide-based microarray platform relies on spotted antibodies which would be used to compare expression levels for various proteins in prostate cancer specimens. Changes in protein levels or activities define cancer at all stages during its development. About 500 antibodies would be spotted on a slide and used to monitor protein levels in various prostate cancer tissues. These would include antibodies against proteins involved in apoptosis, growth, migration, cell division, and so on. Also since most of the proteins orchestrate their effect by changing their state of activation (by phosphorylation or cleavage), antibodies to monitor such changes in activity of the proteins can also be included. This would result in a cohort of proteins, henceforth termed biomarkers, that could then be used to monitor cancer progression. Tissue microarray can also be used to study expression pattern. This is based on spotting hundreds of cancer specimens on a single slide and allow comparative localization of the biomarkers across hundreds of cancer specimens. Finally, as a part of the second aim, we can ascertain the clinical associations of candidate biomarkers using clinically stratified prostate cancer specimens.

Use of Microarray in Water Quality Management

Lyonnaise des Eaux, the world leader in water management, on February 10, 1999 at Paris has entered into a joint research and development (R&D) partnership program with bioMérieux, a world leader in IVD (in vitro diagnostics), to develop GeneChip arrays from Affymetrix that will revolutionize drinking water quality controls. The Lyonnaise des Eaux/bioMérieux initiative is a world-first in the water industry. The endeavor will result in a more efficient, faster and cheaper technology than currently exists for water quality control and will increase consumer safety.

Lyonnaise des Eaux is the international leader in water management, currently supplying 77 million people around the world with drinking water and 52 million people with wastewater services of which 80% are based outside France. Lyonnaise des Eaux controls the entire water cycle including water resources management, drinking water treatment and wastewater treatment. Lyonnaise des Eaux, the water division of Suez Lyonnaise des Eaux Group had a turnover of FRF30.6 billion in 1997.

The GeneChip array which will be applied in this joint R&D partnership program was originally developed by Affymetrix, Inc. (NASDAQ: AFFX). The first commercial applications of the chip were in the biological research fields in 1996. The GeneChip technology is a multidisciplinary system that combines microbiology, photolithography, chemistry, image analysis and bioinformatics.

BioMérieux established a collaboration with Affymetrix, in order to develop new applications for the GeneChip array technology in the fields of bacterial identification and antibiotic resistance, virology, food and cosmetic testing.

Lyonnaise des Eaux and bioMérieux, with a joint scientific team based in France, had invested 8.5 million euros (55 million FRF) in a 5-year R&D partnership program to develop the specific DNA chip. Affymetrix is manufacturing the GeneChip arrays for the partnership.

The Lyonnaise des Eaux and bioMérieux GeneChip array applied to water quality control will offer three main advantages.

Clean water healthy life

- Higher performance than existing technology.
- Time saving in the water control process.
- Significant cost savings.

High performance technology: As seen in medical applications, the DNA chip offers a very accurate screening technology. Applied to water quality controls, it will screen for the presence of microorganisms in the water by matching genetic fingerprints. The DNA chip, compared with existing technologies, allows a wider variety of specific tests to be carried out to determine the quality of the water and will provide the consumers with extra safety.

With the use of molecular biology protocols, the GeneChip array will permit the detection of lower concentrations of microorganisms in the water and the accurate identification of many types of water contaminants.

Time saving technology: The Lyonnaise des Eaux and bioMérieux DNA chip will dramatically cut down on test time (currently 48 hours on average) to just 4 hours. The time saving factor offers 2 major advantages:

- The faster water contamination is detected, the faster relevant water treatment can be carried out.
- The faster normal water services resumes the less penalized the consumers and especially industrial consumers are.

Cost-effective technology: With bioMérieux's expertise and know-how in molecular biology and DNA probes, Affymetrix's GeneChip technology will enable Lyonnaise des Eaux to measure the presence of dozens of microorganisms in one test. With existing technologies, a specific water test is required per microorganism researched, and the standard water microbiology test requires screening for 24 parameters or microorganisms. Currently, the total cost for a regular test amounts to 3,075 euros (20,000 FRF) and is expected to be ten times cheaper with the DNA chip technology.

The DNA chip—a big step forward in technology: The GeneChip technology enables the production of a "biochip" designed to identify fluorescent-labeled DNA or RNA fragments through their hybridization to oligonucleotide probes, which are short DNA sequences. The probes are synthesized on a miniaturized glass substrate, measuring approximately 1 cm^2, on which hybridization occurs. Each hybridization surface may contain a very large number of unique oligonucleotide probes (up to 400,000), permitting several thousand individual nucleotide positions to be characterized at the same time.

A positive hybridization result is detected by the intensity of the fluorescence, which reflects the stability of the hybridization between the oligonucleotide probe and the target sequence of the microorganisms potentially presents in the water samples.

Unlike conventional screening methods based on morphology and physiology characteristics, molecular testing identifies one or several DNA sequences of any microorganisms including bacteria, viruses and parasites. The high resolution DNA chip technology allows a highly reliable and fast screening, and identifying, process for any type of microorganisms in the water.

Scientists from National Research Council of Canada, Biotechnology Research Institute, have designed oligonucleotide-based microarrays which can be used for water-borne pathogen detection.

About the research: Microbial water quality assessment and the rapid detection of pathogenic microorganisms remain great challenges in ecotoxicology, public health maintenance, and, more recently, bioterrorism prevention. The main origin of human water-borne pathogens is the guts of humans and animals. Water contamination events often result from discharges from wastewater treatment facilities, overflowing sanitary sewer systems, waste materials that find their way into domestic and industrial sewage, and runoff of animal fecal matters during storm events. Bacteria within contaminated water represent a highly diversified group. Numerous *Salmonella* spp., *Escherichia* spp., *Shigella* spp., *Yersinia* spp., and *Klebsiella* spp. within the Enterobacteriaceae are human waterborne pathogens found in domestic wastewater. In particular, *Escherichia coli* strains constitute an important set of waterborne pathogens composed of numerous Shiga toxin-producing strains and enterohemorrhagic (e.g. O157: H7) strains associated with potable and recreational water.

Several nucleic acid-based methods have been developed for the rapid detection of pathogens in food, soil, and water with high degrees of sensitivity and specificity and without the need for complex cultivation. In general, these methods allow detection within hours, rather than days as is normally required by culture techniques. Due to its high sensitivity and specificity, PCR is the most commonly employed molecular tool. A major limitation to this approach is the utilization of one specific primer pair per gene detection reaction. Although multiple primer sets may be successfully combined in one reaction, they rarely exceed more than six primer sets due to the generation of nonspecific products or false negatives. Another difficulty with multiplex PCR is that it requires additional postamplification analysis to discriminate the products. Size separation by electrophoresis is frequently used to discriminate multiplex PCR products, but this requires additional labor and that the amplicons of each reaction be significantly different in size, which can limit the primer pairs that can potentially be multiplexed. Consequently, general pathogen detection by PCR can be both labor-intensive and costly.

Microarrays represent an important advance in molecular detection technology, allowing the simultaneous detection of specifically labeled DNAs from many different pathogenic organisms on a small glass slide containing thousands of surface-immobilized DNA probes. Both basic types of microarrays, i.e. immobilized oligonucleotide probes and PCR amplicons, have been used successfully to detect and/or characterize pathogens.

As the sensitivity of microarrays hybridized with total genomic DNA from complex mixtures is usually inadequate to provide detection of low pathogen concentrations, the hybridized DNA (target) usually consists of PCR amplicons. This mode of pathogen detection necessitates the combination of many PCRs prior to their hybridization on microarrays. Scientists used 140 amplicons to characterize 18 pathogenic species, thus constraining the use of microarrays for routine detection of pathogens in wastewater. Target DNA amplification with universal primers to ubiquitous genes prior to microarray hybridization can circumvent this limitation. The cpn 60 gene codes for GroEL, an essential, highly conserved chaperonin protein which displays moderate DNA sequence diversity, making this gene useful in bacterial taxonomy applications.

However, within the Enterobacteriaceae, 16S rRNA and cpn60 sequences may share sufficient similarity to generate cross-hybridization reactions, even when short oligonucleotides are used as probes. As the majority of water pathogens belong to this family, discrimination on the basis of 16S rRNA and cpn60 sequences is challenging. However, sequence diversity within the wecE gene, an Enterobacteriaceae-specific gene which forms part of the wec gene cluster involved in enterobacterial common antigen biosynthesis, has been shown to discriminate among the Enterobacteriaceae most frequently found in water.

The aim of this study was to assess the detection efficacy of an oligonucleotide-based microarray designed with probes specific for the universal targets of 16S rRNA and cpn60 genes in addition to the Enterobacteriaceae-specific wecE genes of several pathogens usually encountered in wastewater.

Summary

A small-oligonucleotide microarray prototype was designed with probes specific for the universal 16S rRNA and cpn60 genes of several pathogens that are usually encountered in wastewaters. In addition to these two targets, wecE-specific oligonucleotide probes were included in the microarray to enhance its discriminating power within the Enterobacteriaceae family. Universal PCR primers were used to amplify variable regions of 16S rRNA, cpn60, and wecE genes directly in *Escherichia coli* and *Salmonella enterica serovar typhimurium* genomic DNA mixtures (binary); *E. coli, S. enterica serovar typhimurium,* and *Yersinia enterocolitica* genomic DNA mixtures (ternary); or wastewater total DNA. Amplified products were fluorescently labeled and hybridized on the prototype chip.

The detection sensitivity for *S. enterica serovar typhimurium* was estimated to be on the order of 0.1% (104 *S. enterica* genomes) of the total DNA for the combination of PCR followed by microarray hybridization. The sensitivity of the prototype could be increased by hybridizing amplicons generated by PCR targeting genes specific for a bacterial subgroup, such as wecE genes, instead of universal taxonomic amplicons. However, there was evidence of PCR bias affecting the detection limits of a given pathogen as increasing amounts of a different pathogen were spiked into the test samples. These results demonstrate the feasibility of using DNA microarrays in the detection of water-borne pathogens within mixed populations but also raise the problem of PCR bias in such experiment.

Application of Microarray in Forensics

Analysis of genetic relatedness is of great importance in forensic casework such as immigration cases involving asylum and family reunification matters. An immigration case often consists of a man who would like to be reunified with his wife and children, and DNA analysis is an important tool in the establishment of the family relations. The cases can be complicated for reasons such as cultural differences in family constellations and complicated family relations giving complex pedigrees.

Investigation of genetic relatedness can also be useful in the identification of deceased individuals, e.g. victims of mass disasters. These cases can be complicated if no first-degree relatives are available, which may be the case in mass disasters where many members of the same family may be among the victims.

There are two types of genetic markers currently used in kinship analysis: short tandem repeats (STRs) and single nucleotide polymorphisms (SNPs). STRs, also known as microsatellites, consist of repetitive DNA (e.g. CGTCGTCGTCGT) where the number of repeats is variable in a population.

SNPs are much more numerous than STRs, but they are not as commonly used in the forensic field, mainly due to their lower degree of variability. As many as eight SNPs are needed to provide the discriminating power of one STR locus. Nevertheless, the interest of SNP markers is increasing because of their many advantages: they have very low mutation rates which is interesting in paternity testing, they can be analysed in short amplicons which is desirable when analysing degraded samples, and finally they are suitable for analysis with high throughput technologies.

However, neither the methods based on STR markers nor the methods for SNP detection currently used in forensic genetics are capable of solving the most complicated immigration and identification cases, since the number of genetic markers is not sufficient.

Difficulties often arise when possible close relationships are to be tested against each other. Examples include issues such as whether a man is the father, uncle or grandfather of a child, or whether two children are siblings or halfsiblings. At present, a maximum of 52 different SNP markers can be detected for forensic purposes. To correctly classify a grandparent–grandchild relation with a probability of 98.8%, 3820 biallelic markers would be needed.

To correctly classify such relations with 99.99% probability, which is the quality requirement of the National Board of Forensic Medicine, as many as 500 000 markers could be necessary. A possible solution to this problem may be the introduction of microarrays, which allow analysis of great numbers of SNPs.

Although hybridization is always the basic mechanism in microarray technology, different methods for SNP detection involving enzymes (e.g. polymerases and ligases) can also be applied using microarrays.

Methods for SNP typing

There are a number of different SNP genotyping techniques available. The majority of those suitable for the microarray format can be assigned to one of three groups based on the molecular mechanism: allele specific oligonucleotide hybridization, primer extension and allele specific oligonucleotide ligation. A previous PCR amplification of the region around the variable site is usually required.

Allele-specific oligonucleotide hybridization

Two DNA targets that differ by one nucleotide can be distinguished by hybridization. Two different probes, specific for the two alleles, are designed (Fig. 11.20).

The variable base is usually in a central position of the probe sequence. To analyse which allele or alleles that are present in a sample, the targets in the sample are allowed to hybridize with the probes. Under optimised conditions, only perfectly matching probe-target hybrids are stable, and hybrids with one-base mismatch are unstable. The hybridization can be fluorescence detected if the SNP-containing targets are fluorescently labelled during PCR amplification. The oligonucleotide probes can be attached to a solid support to create a microarray.

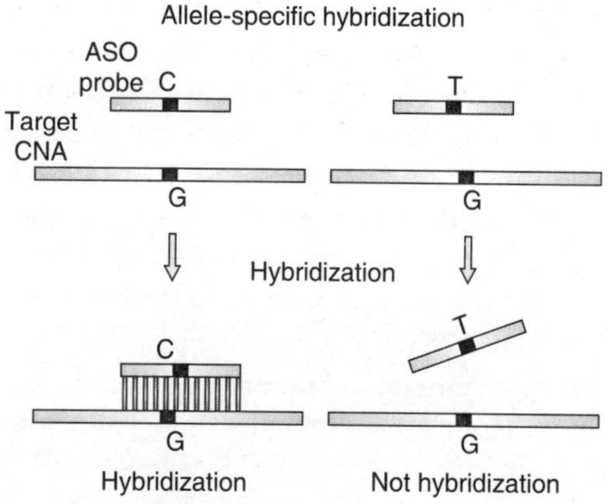

Fig. 11.20: Allele-specific oligonucleotide hybridization. Probes hybridize with the targets containing the SNP. Only perfectly matching hybrids are stable

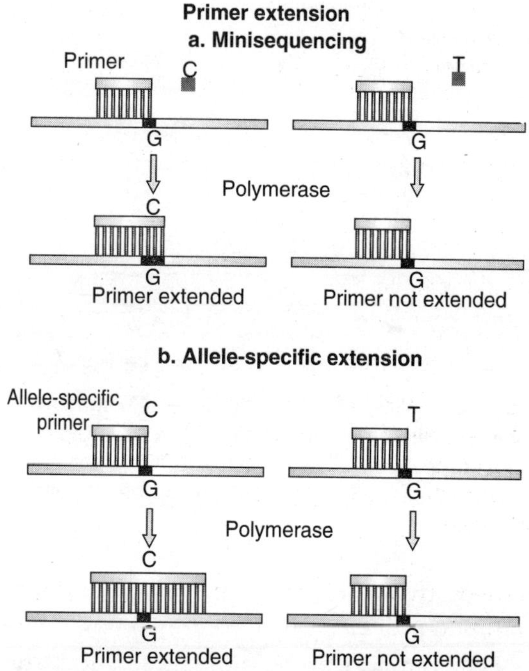

Fig. 11.21: Primer extension. A. Single nucleotide primer extension

Primer extension: DNA polymerase has the ability to incorporate specific nucleotides that are complementary to the sequence of template DNA. There are two main types of reaction: single nucleotide primer extension and allele-specific primer extension.

In single nucleotide primer extension, also called minisequencing, a primer anneals to the target DNA immediately next to the SNP (Fig. 11.21).

The primer is then extended by DNA polymerase with the dideoxynucleotide (ddNTP) that is complementary to the polymorphic site. One way to detect the primer extension is by fluorescence, requiring that the ddNTPs used for extension are fluorescently labeled.

The primers can be attached directly on a chip, in a technique called arrayed primer extension (APEX). In another approach, the primers for the different SNPs have unique tags at the 5′ end. The unique products of the reaction (which takes place in solution) can then be hybridized to specific complementary sequences arrayed on a chip and detected by fluorescence. Another alternative involves detection with MALDI-TOF mass spectrometry, with the matrix on the surface of a chip. The high resolution of mass spectrometry makes it is possible to distinguish which ddNTP has been incorporated.

a. Single Nucleotide primer extension, a primer anneals immediately next to the SNP and is extended with a single nucleotide complementary to the polymorphic base.
b. Allele-specific primer extension. A primer anneals and is extended if there is a perfect match at the SNP site.

In allele-specific primer extension, a primer is extended by DNA polymerase if its 3′ end is perfectly complementary to a DNA template (Fig. 11.21b). Fluorescently labeled nucleotides can be used for the reaction.

Allele-specific oligonucleotide ligation: DNA ligase repairs nicks (i.e. missing phosphodiester bonds) in DNA. Three probes are de-

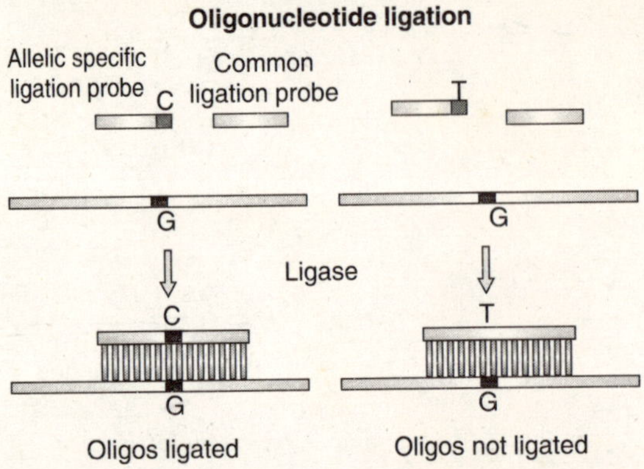

Fig. 11.22: Allele specific oligonucleotide ligation. An allele specific probe and a common probe hybridize adjacent to each other. If there is a perfect match between the target and the specific probe, the probes are ligated

signed: two allele-specific probes and one common probe.

The common probe anneals to the target DNA immediately after the SNP. The allele-specific probes compete to anneal next to the common probe, resulting in a double stranded region containing a nick at the SNP site. The allele-specific probe will only be ligated to the common probe by DNA ligase if it matches the target DNA perfectly. There are different methods for detecting the ligated product, e.g. labeling with different fluorescent dyes. The ligation products can contain tag sequences that can hybridize to probes on a chip.

Commercially available microarrays

There are a great number of DNA microarrays available in the market, and the development in the field is rapid. A selection of microarrays suitable for SNP analysis will here be presented.

Affymetrix: Affymetrix provides five different GeneChip arrays with fixed sets of probes and with analysis capacities ranging from 10000 to more than 906 600 SNPs (GeneChip Arrays). They also have custom design alternatives. Arrays, reagents, instrumentation (hybridization oven, fluidics system and scanner) and data analysis tools have been developed as a single platform. The underlying reaction principle is allele-specific oligonucleotide hybridization and the detection is based on fluorescence (GeneChip Technology). The arrays are manufactured using light-activated oligonucleotide synthesis with a physical mask.

The Affymetrix microarrays have been used in several research studies, for example the GeneChip Mapping 100 K Set was recently used for SNP genotyping to identify genes associated with asthma in Puerto Ricans. The study provided evidence that chromosome region 5q23 is associated with asthma.

Asper Biotech: Asper Biotech produces microarrays consisting of oligonucleotides spotted on glass slides, as a part of the Genorama Genotyping Platform (Genorama Geno5 typing Platform). The platform also includes a four colour detector and software. Some established and validated tests exist, but according to the manufacturer, it is easy and affordable to establish custom tests, where the design, development and validation are performed by Asper Biotech. The reaction principle is single nucleotide primer extension, where the primers are attached to the chip (i.e. APEX technology). The terminator nucleotides are labeled with different fluorophores and detection is based on total internal reflection fluorescence (TIRF). A microarray from Asper Biotech has been evaluated for the detection of TP53 mutations in breast and ovarian cancers. TP53 is a tumour suppressor gene, and mutations in the gene are associated with many different cancers. It was shown that the TP53 microarray was a fast and reliable method to screen for mutations in tumours.

Beckman Coulter: Beckman Coulter sells a microarray system called GenomeLab SNP stream Genotyping System. It includes microarray plates with 384 wells where each well can detect 12 or 48 SNPs, resulting in a capacity of analysing roughly 4600–18400 SNPs per plate. The method is based on single nucleotide primer extension followed by hybridization to the microarray plates. The oligonucleotides on the arrays are complementary to tags of one of the 12 or 48 extension primers. The system includes reagents, a two colour imager and software developed to work together (GenomeLab SNPstream Genotyping Series). The SNPstream Genotyping System has been used in a test for blood donors, where SNPs linked to 22 different minor blood antigens are simultaneously tested. The test could be used by blood banks to decrease the costs of finding matching donors.

CombiMatrix: CombiMatrix manufactures CustomArray chips that can be used for SNP detection (CustomArray—Overview and Technology). The oligonucleotide probes can be designed by the user, with help from an open source CombiMatrix probe design system. The probes are then synthesised directly on the chip with a special method called electrochemical detritylation. There are two different protocols for SNP detection; one is based on allele-specific oligonucleotide hybridization (CombiMatrix re-sequencing and SNP detection), whereas the other uses a combination of hybridization, extension and ligation (CombiMatrix Hyb & Seq Protocol). The products can be detected with fluorescence (e.g. Cy3 and Cy5). One microarray can analyze up to 3000 SNPs. The arrays are compatible with many different standard fluorescent microarray scanners; CombiMatrix recommends Genepix 4000B and 4200A from Molecular Devices and ScanArray 4000, 5000, Lite and Express from Perkin Elmer (CustomArray—Overview and Technology). The arrays are also compatible with common sample preparation technologies and hybridization buffers. Software for extracting microarray data is included in the cost of the arrays. In some assays, each microarray can be used four times since target DNA can be removed with the CustomArray Stripping Kit.

CustomArrays from CombiMatrix have been used to study sludge bioreactors enriched with the bacterial species Candidatus Accumulibacter phosphates (CAP). The presence of RNA from 1297 bacterial and 1440 phage genes in sludge samples was tested through hybridization with probes on the arrays. It was shown that highly related but geographically remote CAP strains differ in genomic regions encoding phage defence mechanisms, i.e. they must adapt to the local phage predation pressure.

Febit: Febit provides a custom design microarray technology for SNP typing based on allele-specific oligonucleotide hybridization (SNP typing and resequencing). The platform, called Geniom, is special in the sense that microarray design, production and analysis can be performed with the same instrument (Geniom Genotyping). Hence, the microarrays can be designed by the user, supported by the associated software. The oligonucleotide probes are synthesised on the chips by a light-activated process. The resulting chips are called biochips, and one chip contains eight independent arrays with 6500 spots each. 4–24 spots are needed to characterise one SNP, resulting in a capacity to analyse roughly 2000–13 000 SNPs per biochip. The standard detection is based on biotin/SAPE (streptavidin-R-phycoerythrin) but filter switching makes standard fluorescence detection possible as well. The Geniom system has been used to analyse transcription levels of 3278 genes (covered by 24 125 probes) in the filamentous fungi species *Aspergillus nidulans*. The species is used as a model organism for fungi, and this study compared the wildtype with a mutant (creA_4) using either glucose or ethanol as a carbon source. Through hybridization of RNA to the arrays, it was shown that repression of genes for carbon catabolism in the fungi is dependent both on the presence or absence of the mutation, and on the carbon source.

Data analysis: Software for converting the fluorescence image (or similar) into sequence data is usually provided by each company, and is often included in the cost of the arrays. To determine genetic relationships additional software would be needed. The freely available programs Merlin and Familias and the commercially available DNA-view may be useful for this purpose. They can all be used to calculate probabilities for different relationship hypotheses and can handle both STR and SNP data. However, at present Merlin is the only one that handles linked markers, which would be necessary as the risk for linkage increases when many markers are used. To calculate probabilities for kinship relations, frequencies of different SNP alleles in a population will be required. Such population data can be found on the SNP website of the National Center for Biotechnology Information (NCBI Entrez SNP).

Practical application: Illinois State Police Research & Development Center & Biochip Technology Center, Argonne National Lab have developed MAGIChip™ (Microarrays of gel immobilized compounds on a chip) which would very useful in forensics.

Microarray in Genetic Disease Monitoring and Toxicogenomics

Scientists at the National Human Genome Research Institute (NHGRI) at The National Institutes of Health (NIH) have developed a new genetic test that, for the first time, can easily distinguish between hereditary and sporadic forms of breast cancer. This new approach should make it possible for physicians to quickly and accurately diagnose the cause of an individual woman's disease and may ultimately guide decisions about the most effective treatment.

The international team of researchers, led by Dr Jeffrey Trent, NHGRI's scientific director and head of the NHGRI Cancer Genetics laboratory, used a new technique called gene-expression profiling to differentiate between breast tumors that were caused by inherited genetic changes and those that were not. Based on a laboratory method that uses a type of DNA chip called a microarray, the researchers simultaneously assessed how active some 6,000 genes within breast cancer cells were, altogether making nearly 250,000 measurements. The study revealed clear differences in the patterns of gene activity in breast tumors, patterns that can be as unique as a fingerprint, pinpointing into which group a woman's cancer belongs.

Genetics of breast cancer: Earlier insights tended to give a fairly static view of the genetic changes that produce a tumor. The current study demonstrates the power of a dynamic analysis that shows the interaction of many genes within a cell's metabolic pathways. In what may become a new research and diagnostic trend, the team has established that it can differentiate between hereditary and non-hereditary breast cancer by studying thousands of genes simultaneously.

Physicians have long known that breast cancer can run in families. Approximately 5–10% of breast tumors are hereditary; the remaining cases are caused by genetic changes that occur during a woman's life and are commonly called sporadic. In the mid-1990s, scientists identified mutations in genes now called BRCA1 and BRCA2 that are the major cause of the hereditary form of the disease. Women inheriting these mutations have a 40 to 85% lifetime risk of developing breast cancer, as well as an increased risk of ovarian cancer.

Telling the tumor types apart with traditional techniques is not easy. When we look at these tumors under a microscope, based on their shapes and other features, it is very difficult to tell which tumors are caused by BRCA1, and virtually impossible to distinguish cancer caused by BRCA2 from those caused by non-inherited mutations. Furthermore, the BRCA1 and BRCA2 genes are very large, and searching through them for mutations is both complicated and expensive.

To determine whether gene profiling could tell the difference between sporadic tumors and those with BRCA1 or BRCA2 mutations, the research team examined samples of tumors that had been surgically removed from 22 breast cancer patients at the University of Lund in Lund, Sweden. Fifteen of the women were known to have hereditary breast cancer based on family history and other analyses. Of the 15, seven had mutations in BRCA1; eight had mutations in BRCA2. The control group was comprised of eight women with sporadic cases that had no evidence of any family history of the disease.

When the team examined the gene-expression profiles for the 22 patients using the microarray analyses, they were able to quickly and accurately differentiate the BRCA1 from the BRCA2 inherited changes as well as from the non-inherited genetic changes.

Microarray analysis: The team's approach relied on a cutting edge technology called a DNA microarray, or sometimes, a gene chip. The microarray itself is merely a glass slide with tiny dots of DNA from different genes arranged in a grid-like array. Using microarrays, the activity of thousands of known genes can be quickly tested in a cell sample.

To perform the analysis, scientists isolate a special collection of molecules from the test cells called messenger RNA or mRNA. These molecules are produced by active genes and indicate which of the estimated 30,000 genes are "turned on" in a cell. The information in the messenger RNAs can be converted into a form of DNA called complementary DNA or cDNA—kind of like copying a music CD to digital audiotape; same information, just in a different form. A fluorescent label can be added to the cDNA made from the test cells. A robot then systematically deposits a sample of the cDNA onto the glass slide. If the sample contains cDNA that matches the a particular gene on the microarray, the cDNA sticks to that spot on the array, like two pieces of Velcro coming together. Unmatched cDNA is washed away.

Automated computerized detectors then measure the amount of fluorescence for each spot. The brighter the fluorescence, the more cDNA has attached to the known gene, and the more active the gene must be in that tissue sample. The result appears as a pattern of spots that are either bright or dark, showing whether the known gene is active or inactive in the tumor cell.

Because of the number of gene samples, and the number of patients tested, the study produced almost a quarter million data points. To handle the huge amounts of information generated by this process, the researchers turned to computer-based statistical analyses. With the help of statisticians from the NIH's National Cancer Institute, Agilent Laboratories, and Texas A & M University, the researchers were able to show that the differences between the groups of genes from the cancer cells were, indeed, statistically significant.

Use of Microarray in Molecular Diagnostics

Molecular diagnostic tests typically analyze key protein, DNA, or RNA markers to characterize diseases. Immunohistochemistry, western blotting, DNA sequencing, polymerase chain reaction (PCR), southern blotting, northern blotting, and quantitative reverse-transcription PCR (qRT-PCR) are molecular techniques that have been effectively used in clinical diagnostic laboratories, where broadly trained technologists perform the tests rather than researchers with specific methodological expertise. However, more-parallel molecular tests that involve simultaneous analysis of tens to hundreds of markers within a clinical sample will soon be required to allow maximum translation of postgenomic research to patient care.

Many basic research projects have focused on methods to measure the behavior of the approximately 30,000 genes in the human genome. This research has been done at different levels within the gene, measuring both genotypic and phenotypic changes. For example, parallel analysis has been applied at the DNA level to analyze sequence polymorphisms that distinguish closely related human alleles or microbial genes; at the level of DNA methylation patterns; at the mRNA level to analyze gene expression patterns; and at the protein level.

With the development of microarray-based tests for diagnosing a wide range of diseases, the ability to gain more information from a limited clinical sample by using highly parallel expression–analysis techniques has emerged. This technology's greatest potential exists for those diseases in which histological differentiation between disease entities is difficult and the consequences of inappropriate treatment can be serious. In the research environment, analyzing sets of gene products with microarrays, rather than individual proteins or RNAs, has been shown to provide effective diagnoses. However, improving high-throughput diagnostic methods will require the continued development of new second-generation custom microarrays. Next section describes the challenges in developing this technology and the custom microarray techniques that are currently available.

Microarrays as transcription analysis tools: High-density DNA (HD DNA) microarrays have been useful in the semi-quantitative, or comparative, analysis of thousands of individual messenger RNA (mRNA) species. Experiments using HD DNA microarrays suggest that new tests measuring the expression of tens to hundreds of genes will be better able to characterize disease states than is currently possible using histopathology and other existing molecular methods.

As stated above, clinical application of this DNA microarray-based research will require further development of sophisticated new second-generation custom DNA microarray platforms that are more suitable for routine clinical laboratory testing.

HD DNA microarrays are small devices that hold tens of thousands of gene capture probes at known positions on a solid surface. One of the most widely used platforms is the arrays by Affymetrix Inc. (Santa Clara, CA), in which DNA oligonucleotides are chemically synthesized in situ on the array surface using masked photolithography. Another widely used platform is the cDNA array, in which PCR products are generated separately and placed on the array surface using a robotic arrayer. Thousands of laboratories have used these two platforms to catalog the expression of large numbers of genes in normal and diseased cells and tissues.

When transitioning from a normal to a diseased state, cells and tissues undergo a variety of histological and biochemical changes. These changes reflect in large part the dysregulation in the expression of genes. The quantities and types of proteins that are expressed from their respective genes determine the normal and abnormal functions of cells. When genes are expressed, their DNA sequences are first copied into temporary molecules, or mRNAs. These mRNAs are then translated into proteins that perform the functions and generate the structures that allow cells to conduct their biological activities.

Researchers can take advantage of this flow of genetic information from DNA to mRNAs to proteins by purifying populations of all mRNAs expressed in specimens, synthesizing fluorescently labeled cDNA copies of the mRNAs in vitro, and hybridizing the labeled cDNAs to a microarray. The intensity of the fluorescent signal at each gene-specific probe on the microarray reflects the level of mRNA expressed from that gene in the specimen. By measuring the types and quantities of tens of thousands of individual mRNAs expressed in samples, researchers can infer information regarding the state of gene expression, the approximate quantities and types of proteins present, and the biological behavior of the sample.

For several decades, alternative methods have been available for identifying genes whose expression patterns are altered in disease. However, HD DNA microarrays have increased research scientists productivity in their gene-hunting efforts. As little as five years ago, identifying even a few differentially expressed genes may have taken several years and cost tens of thousands of dollars. Today, HD DNA microarrays can identify 10 times that number of genes in a few months and at a tenth of the cost.

Advances in lymphoma diagnostics: Recent studies on the biology of non-Hodgkin's lymphomas (NHL) illustrate some of the diagnostic advances made possible by HD DNA microarrays. In 2000, it is estimated that 54,900 patients were newly diagnosed with NHLs and 22,553 died of these diseases. Although there are many NHL subtypes, the diffuse large B cell lymphoma (DLBCL) subtype is the most common, comprising 30% of all cases.

Despite the complexity of this classification system, it is increasingly recognized that the NHL categories themselves encompass multiple disease subtypes with distinct molecular defects and clinical outcomes. For example, while 35–40% of patients diagnosed with DLBCL can be cured with chemotherapy, the remaining 60–65% die of this disease. Numerous attempts to distinguish patients who are likely to respond to chemotherapy from those unlikely to respond have been unsuccessful.

In addition, these DLBCL tumors are indistinguishable using classical histopathology and currently available molecular markers.

Researchers have approached this problem by using HD cDNA microarrays to profile gene expressions in DLBCL. These researchers identified genes that classified DLBCL specimens into distinct subclasses. One subclass, germinal-center B cell-like DLBCL, showed an expression pattern similar to mature germinal-center B cells and portended a more favorable prognosis with a 60% five-year survival rate. On the contrary, two other DLBCL subclasses, activated B cell-like DLBCL and type-3 DLBCL, showed much less favorable prognoses with a 35% five-year survival rate.

Microarrays as Genotyping Tools for Infectious Diseases

Despite the abundance of protein diagnostics, diagnosing diseases by examining the qualitative presence of specific nucleic acid sequences has been adopted for infectious diseases. While the amplification of gene sequences through PCR or other similar methods can indicate the presence or absence of a pathogen genomic sequence, nucleic acid hybridization is almost always necessary to confirm that the amplification is sequence-specific. Suitably designed microarrays are versatile tools for this function since the hybridization data from an array can identify the pathogen genomic sequence in a more rapid and cost-effective manner than DNA sequencing.

The sudden emergence of new strains of pathogens has brought the need for rapid nucleic acid diagnostic tools to the forefront. Although customizable nucleic acid testing systems for the rapid development of diagnostics are still new and emerging, the urgent medical need for high-speed diagnostics is steadily increasing. For example, the recent outbreak of severe acute respiratory syndrome (SARS) and its rapid spread left no time for generating an antibody-based diagnostic test.

Initial identification of the virus was based on symptom presentation and high-resolution electron microscopy of sputum samples. However, gene sequencing of the emergent SARS strains permitted rapid identification of the virus by using a nucleic acid amplification and hybridization/sequencing approach.

Monitoring how the genotype of viruses evolves is critical for understanding the pathology of a disease as it spreads throughout a population. Viruses such as influenza, poliovirus, rhinoviruses, and HIV are known to have high rates of mutation, in some cases even causing attenuated strains that are developed for vaccines to revert to viruses with a wild-type phenotype. Although it remains possible to generate antibodies for all of these pathogen strains, identifying strains rapidly is more easily achieved using microarray tools developed specifically for this purpose.

Microarray in Food and Agricultural Biotechnology

Development of a DNA microarray supporting both functional and identification probes for dairy (food) microorganisms: Oligonucleotide probes targeted on microbial functional genes will be designed on genetic regions encoding activities that influence the ripening process and flavour formation in food products (hydrolytic enzymes like peptidases, esterases and glucosidases, amino acid transaminases, permeases, etc.). Such genes will be selected for the most important components of food microbial flora belonging to different genera and species. Species-specific oligonucleotides, targeted on taxonomically discriminating genetic regions, probes will be designed for the identification of dominant species in the selected products. Finally a DNA microarray support, inclusive of all the oligonucleotides selected, will be created that will enable, at a time, the identification and functional characterisation of single microbial strains.

Microarray is also used in analysis of expression of transgenic plant and effect of microorganism and stress on plants.

Microarray in Drug Safety

cDNA microarray technology, which can be used to analyze changes in genome-wide patterns of gene expression is one new methodological advance that may revolutionize the way some toxicological problems are investigated. The application of a large number of genes or expressed sequence tags in a condensed array on glass slides or nylon filters comprises a cDNA microarray. Alternatively, specific oligonucleotides that are complementary to known genes or expressed sequence tags are deposited on a miniature matrix by a photolithographic process to create an oligonucleotide-based microarray. Either cDNA microarrays or oligonucleotide-based chips may be used for gene expression analysis. Oligonucleotide-based DNA chips are also used for analyzing sequence variations in genomic DNA for screening individuals for DNA mutations and polymorphism variations. This approach has been recently reviewed.

Changes in gene expression in a tissue may result from differences in physiology, developmental stage, pathology, or environmental exposure. These changes can now be measured using cDNA or oligonucleotide-based microarrays, which are used to compare directly the gene expression profiles of two RNA samples that are simultaneously hybridized to the chip. The potential analysis of the expression of thousands of genes in one experiment now allows investigators to consider addressing some important biological questions that have not been easily addressed with traditional expression-based technologies, such as northern blots, in situ hybridization, or RNase protection assays, which examine gene expression changes of only a few genes at a time. The ability to examine thousands of genes (potentially all of the genes in a given cell type) provides new insights into the effects of chemical or drugs on biological systems. Microarray technology will be useful to identify toxic substances individually or in mixtures, to determine whether toxic effects occur at low doses, and to extrapolate effects from one species to another. Assuming that exposures to different classes of toxicants result in distinct patterns of altered gene expression, in addition to common changes associated with the subsequent toxic response, microarray technology can be used to categorize and classify these effects through the direct comparison of gene expression signatures in exposed and control samples.

One example of the use of cDNA microarrays is in the process of drug development. Given that advances in genomics and combinatorial chemistry are leading to the discovery of many new potential drugs, surrogate markers of efficacy and safety are needed to expedite clinical trials. Gene expression profiles can be used as a proof of principle assay to show an effect of a candidate drug in vivo. Furthermore, cDNA microarrays can be used to detect toxic responses in target and nontarget tissues in rodents and humans. The dose of a drug that maximizes the therapeutic index can potentially be determined from such measurements, which will improve optimization of lead compound development. By the use of cDNA microarrays, toxic or unanticipated responses in humans may be determined early in a clinical trial prior to overt tissue toxicity, providing a rapid, sensitive surrogate of safety, which is essential for improved clinical trials. Also, microarrays may help identify susceptible individuals who respond to a treatment or who exhibit adverse effects to drugs.

Government Support for Biochips in India

TIFAC is extending its support for projects in Microarray under its Bioprocess and Bioproducts programme.

TIFAC

Technology Information, Forecasting and Assessment Council (TIFAC) is an autonomous organization set up in 1988 under the Department of Science & Technology to look ahead in technologies, assess the technology trajectories, and support technology innovation by network actions in select technology areas of national importance. In 1993, TIFAC embarked upon the major task of formulating a Technology Vision for the country in various emerging technology areas. The outcome of the Technology Vision 2020 exercise led to the set of 17 documents, including sixteen technology areas and one on services.

Bioprocess and Bioproducts Programme of TIFAC was launched in January 2007 with the aim to address critical technology needs for biotransformation and enzymatic processes towards development of active pharmaceutical ingredients, neutra-ceuticals, phyto-chemicals, value-added bio-products, bio-energy and bio-fuels. It also has special provisions for funding in microarray technology.

DBT

The setting up of a separate Department of Biotechnology (DBT), under the Ministry of Science and Technology in 1986 gave a new impetus to the development of the field of modern biology, biotechnology and microarray in India. In more than a decade of its existence, the department has promoted and accelerated the pace of development of biotechnology in the country. Through several R&D projects, demonstrations and creation of infrastructural facilities a clear visible impact of this field has been seen. The department has made significant achievements in the growth and application of biotechnology in the broad areas of agriculture, health care, animal sciences, environment, and industry.

A unique feature of the department has been the deep involvement of the scientific community of the country through a number of technical task forces, advisory committees and individual experts in identification, formulation, implementation and monitoring of various programmes and activities.

In the proposed programmes for the 10th plan, it aims at following collaborations for development of microarray technology along with several other thrust areas in Biotechnology.

Establishment of multilateral/bilateral Institutions/programmes in association with Ministry of external affairs under consideration are:

- Joint centres in biotechnology specially, with Germany and France.
- Indo-ASEAN Institute of Biotechnology in Jakarta.
- Indo-ASEAN Biotechnology Network.
- India-Singapore Joint Biotechnology park

DBT Has been working towards creation of microarray facilities for defining the expression and functions of genes. For important crops like rice, wheat, brassica, chickpea, a map-based marker assisted technology development for precision breeding, as well as gene identification through in situ molecular hybridization.

Major scientific achievements: DBT Centre of excellence for development of high throughput approaches to understand molecular basis of heterosis in rice for precision breeding at University of Delhi South Campus, New Delhi in collaboration with Maharashtra Hybrid Seed Company Ltd., Jalna, Maharashtra.

The goals of the centre are to develop an understanding of the molecular basis of heterosis by microarray-based transcriptome profiling of Paddy hybrid 5401: female, PMS 79: Restorer: PR 319 and its comparison with Hybrid, RH10: Female, PMS 6A: Restorer, PR 78 at selected stages and to select novel genes and promoters with potential to contribute to heterosis and their evaluation in transgenic rice for yield improvement. It is also planned to compile logistics for developing a chip/PCR-based diagnostic system for the selection of suitable parents for generation of hybrids with desirable traits.

Under the programme on cancer biology and therapeutics at IISC Bangalore, efforts have been initiated to identify gene expression signatures in various forms of breast cancers and breast cancer stem cells. About 96 breast cancer tissues and 82 normal breast tissue have been collected for microarray analysis. In order to elucidate the role of AP2 a in the progression of breast cancer, a total of 314 breast cancer cases (retrospective) along with the respective survival data of various clinical forms were collected. Immunohistochemical staining for all these samples have been completed. Towards developing cancer cytotoxic drugs based on metal complexed compounds, two series of ruthenium and platinum complexes were prepared. The cytotoxicity of the complexes and the corresponding ligands has been established by MTT assay on cancer and normal cells.

Under the Rice Functional Genomics Programme, an elaborate transcriptome profiling for most of the developmental stages of panicle and seed development has been completed by using the rice GeneChip microarrays and SAGE. For a few target genes, overexpression, silencing and promoter deletion constructs have been prepared and the work on their functional validation initiated.

Microarray setup by DBT:

1. Centre for DNA Fingerprinting and Diagnostics (CDFD), Hyderabad.
2. National Brain Research Center (NBRC) Gurgaon.

Support for governments of other countries towards microarray: US Airforce has funded $ 890,000 TO CMBX an Australian company to develop Fluchip for identifying all Influenza strains.

Microarray Quality Control Project (MAQC) has been initiated by US government which has been supporting research in microarray.

Japan Science & Technology agency has extended its support for microarray research.

Canadian government support: The Centre for Applied Genomics is a genome centre in the Research Institute of The Hospital for Sick Children, and is affiliated with the University of Toronto. TCAG also operates as a Science and Technology Platform of Genome Canada, with an emphasis on next-generation sequencing (NGS) and bioinformatics support. Research at TCAG focuses on the genetic and genomic basis of human variability, health and disease, including research on the genetics of autism spectrum disorders and structural variation of the human genome. The Centre is located in the MaRS Discovery District in downtown Toronto, Canada.

Microarray and gene expression: The Microarray and Gene Expression Core Facility has a dedicated manager, and operates Affymetrix and Agilent technologies. All Illumina technologies are contained within the Genetic Analysis Core Facility. Affymetrix MegAllele and Single Nucleotide Polymorphism (SNP) microarrays, Affymetrix gene expression, tiling, promoter and exon microarrays, and Agilent oligonucleotide microarrays are all run. Additionally, there are a wide variety of analytical software packages available for onsite data analysis (Figs 11.23a and b).

(a) (b)

Figs 11.23a and b: The Centre for Applied Genomics, Canada

Resources

- Collaborative research initiatives
- Microarray technology development
- Bioinformatics support and custom data analysis tools

Patent Issues in Microarray

Introduction

The drug discovery process seeks to develop a biological or chemical entity that, when administered to a patient, will improve disease symptomatology or actually treat the underlying pathophysiological basis of a particular disease state. A target is defined as the biological entity, usually a gene, mRNA or protein, with which a pharmaceutical interacts. The cost of developing therapeutics has increased considerably in recent decades. US-based companies are likely to spend over US$60 billion by 2005, and a minimum of 25% of all pharmaceutical sales revenue will be invested in research and development. Currently, the average cost of bringing a new drug to market is US$750 million and this involves 10–12 years of arduous development and clinical trials. The success of a pharmaceutical company is therefore dependent on the overall efficiency of this process and an accelerated development time. Bioinformatics and microarray technologies are an integral component of this process. In 2000, the global bioinformatics market was ~ US$350 million, and by 2005 it is projected to be US$1045 million. Bioinformatic approaches coupled with DNA microarray data are uncovering novel biological mechanisms and predicting the effects of drug treatments. They are a necessary prelude to the expensive and time-consuming preclinical and clinical tests.

After genome sequencing, data derived from DNA microarray approaches has become the source of genomic information most widely used by the pharmaceutical industry. Consequently, the market for DNA microarrays is expected to increase at a rate of 33% a year.

The DNA microarray market is expected to double in the next 5 years. This expansion will encompass the areas of robotics, slide and attachment chemistries, labeling and detection strategies, and informatics and data management systems. This growth is not surprising given the powerful nature of this technology and its application to understanding the genomic basis of disease.

Patentable subject matter: Technological developments can be protected either in the form of a trade secret indefinitely by refraining to disclose the details of the invention, or in the form of a patent where in exchange for exclusively utilizing the invention for a defined period of time (20 years) details of the technology are comprehensively described to the public. In order to be awarded a patent, the invention must have utility and be sufficiently novel so that it is not an obvious

Fig. 11.24: Picture of European patent office

variation on something that has already been created. The US has a history of patent law dating back to 1790, with the first patent laws enacted shortly after the Constitution was drafted. In Europe, France was the first nation to adopt patent laws that were established in 1791. Currently, in the US, patent applications are reviewed at the United States Patent & Trademark Office (USPTO) in Arlington, Virginia.

In Europe, the European Patent Convention was signed in 1973 in order to harmonize patent laws amongst the various member states, and subsequently the European Patent Office (EPO) was established with is primary location in Munich, Germany. Generally, both US and European patent laws are complementary, with the exception of obtaining priority rights. In the US, the inventor claims priority on the date of conceiving the invention whereas in Europe the priority date is the date of filing the patent application.

Drug discovery research is in many ways a unique field in that in the course of daily experimentation, scientists typically perform processes using technology or reagents that are protected by patents. Many biological patents have been issued to protect these processes, the most controversial in recent times being PCR. License, royalties and patent fees are paid to companies and universities, which help to recoup the costs incurred in the development of the particular technology. This maintains an active research and development stream, ensuring the viability of technology innovation. Patenting of inventions in proteomic and genomic fields has been prolific in the last few years.

Requirements for patentability: In order for an invention to be patented, it must be novel and useful (a concept known as utility) and must not have been obvious 'to one of ordinary skill in the art'. Furthermore a patent application must comply with a series of standard requirements, namely a written description, enablement and best mode. The invention must be new, believable and possess real world use. Although a product or process may be novel, the invention may not obtain patent protection if the differences between the invention and the prior art are such that the invention as a whole would have been obvious at the time the invention was made. A patent application must contain a written description of the particular invention sufficient to enable any person skilled in the art to which it pertains to make and use the invention. The specification must disclose the best way the inventor knows of making the invention and using it at the time the application was filed. The test for satisfying the best mode requirement is subjective, as it depends entirely on what was in the inventor's mind at the time of filing the application, regardless of the knowledge or skill level.

The early development of microarray technology led the leaders in the market to compete amongst themselves in establishing their respective IP space. In order to understand to what extent these companies' patent portfolios cover, one needs to look at the legal requirements for obtaining a patent. Generally in order for an invention to be considered patentable in the US, it must meet the requirements of 35 United States Code (USC) Sections 101, 102 and 103. In Europe, the requirements stated in European Patent Convention (EPC) Articles 52, 54 and 56. 35 USC Section 101 and EPC Article 52 require that an invention be new and useful. Section 102 and Article 54 require that it must be novel. Section 103 and Article 56 refer to the requirement that an invention must be nonobvious, which means that differences between the subject matter sought to be patented and prior art are such that the subject matter as a whole would have been obvious at the time the invention was made to a person having ordinary skill in the art. Further, in order for an issued patent to be valid, the inventor must describe the invention so that a person of ordinary skill in the art can make or use it as required by 35 USC. Section 112, paragraph 1. The invention must also be described in certain and definite terms under Article 35 USC Section 112, paragraph 2. Under EPC Article 83, a patent application must disclose the invention in a sufficiently clear manner, allowing it to be reproduced by a person skilled in the art.

Microarray patents issued and their enforcement: DNA analysis methods are inherently ligand assays that rely on the binding of target molecules to a specific recognition reagent. DNA analysis techniques thus frequently employ approaches historically exploited in an immunoassay context. Two approaches employed in immunoassays are competitive (reagent limited) and immunometric (reagent excess). The basic principle of a competitive immunoassay is that a constant amount of labeled antigen competes with the unlabeled antigen in a specimen (serum, plasma or cell lysate) for a limited number of antibody binding sites. The amount of labeled antigen bound to the antibody is thus inversely proportional to the amount of unlabeled antigen in the specimen. The immunometric assay such as a 'two-site' sandwich assay is based on the formation of a sandwich complex between the capture antibody, the antigen and the labeled antibody. The capture antibody and the labeled antibody bind two distinct epitopes on the antigen. The amount of labeled antibody bound to the antigen is thus directly proportional to the concentration of the antigen in the specimen. The 'ambient analyte immunoassay' pioneered by Ekins and colleagues, has been utilized in the development of multi-analyte immunoassays, enabling the multiplexing of hundreds of analytes from a single specimen. In 1989, Ekins described a ratiometric immunoassay utilizing antibodies conjugated to two separate fluorescent dyes. Immunoassays have relied on antibodies attached to solid supports, an analogous approach being later applied to DNA measurement. Microarrays were initially developed for immunodiagnostic purposes because immunoassay constituted a methodology of unique biomedical importance in the 1980s.

In this same year that Ekins published his seminal paper, Abbott Laboratories was issued a patent describing a system that precisely microdispenses chemical reagents in defined regions for the purposes of developing chemical diagnostic assays. However, it was not until the early 1990s that the first patents covering DNA microarrays were issued. Hyseq (Sunnyvale, CA,USA), Affymetrix (Affymax) (Santa Clara, CA, USA), Oxford Gene Technologies (Oxford, UK) and Stanford University (CA, USA) received patents that covered microarray manufacturing, experimental processing and genomic profiling. Difficulties in understanding the extent of these various patents led to a series of infringement lawsuits, as these parties sought to define their IP space.

Hyseq and Affymetrix: Hyseq issued patents US5202231, US5695940 and US5525464; these patents issued over a 4-year period cover a method of sequencing DNA by hybridization on an array. This method involves using oligonucleotide probes with lengths between 11–20 nucleotides that are hybridized to the target nucleic acid sequence. These complementary oligonucleotide probe sequences are overlapping in sequence thereby enabling identification of the target nucleic acid sequence. This method can discriminate perfect hybrids from hybrids containing a single nucleotide mismatch thereby enabling accurate DNA sequencing in a high-throughput array format. Affymetrix issued many early patents regarding the microarray manufacturing process. Patents US5445934, US5744305 and US5677195 claim methods of manufacturing polymers on solid supports using light-directed spatially parallel chemical synthesis.

These manufacturing techniques utilize Very Large Scale Immobilized Polymer Synthesis (VLSIPS™) substrate technologies that can be applied for the synthesis of peptides and oligonucleotides on solid supports. The manufacturing process utilizes photoremoveable groups attached to solid supports. Upon exposure to light, selected areas become activated and permissible to attachment to monomers such as nucleotides and amino acids. Patent US5744305 relates to arrays of polynucleotides at densities > 400 polynucleotides/cm^2.

Affymetrix also issued patents regarding the array experimental applications and analysis. US5800992 and US5871928 claim different applications for which arrays are used. These applications include DNA sequencing, DNA fingerprinting, chromosomal mapping and specific interaction screening. The US5800992 patent claims analyzing two different samples simultaneously using different fluorescent colors. US5795716 claims a computer system that identifies unknown sequences by analyzing the fluorescent intensities of hybridized nucleic acid probes. It also claims using a set of match and mismatch oligonucleotide probes to determine the identification of a given sequence. In addition, this patent describes a means of analyzing the image of the hybridized chip to identify sample sequences.

In March 1997, Hyseq filed a lawsuit against Affymetrix for infringement of Hyseq's US5202231 and US5525464 patents, and in December 1997, Hyseq filed another lawsuit for infringement of US5695940. Affymetrix responded to these claims, stating that the US5795716 patent represented a novel technique because it employed a set of matched and mismatched oligonucleotide probes to identify DNA sequences whereas the Hyseq technology utilizes overlapping perfectly matched probes. Additionally, Affymetrix alleged that Hyseq infringed on the US5744305 patent regarding array densities greater than 400 polynucleotides/cm^2.

In January 2001, the US District Court of the Northern District of California issued a Markman ruling, which is used to provide clarity to disputed patent claims. This ruling established that Affymetrix patents US5795716 and US5744305 were valid and rejected all of Hyseq's arguments regarding infringement by Affymetrix. Affymetrix and Hyseq arranged a settlement in October 2001 that dismissed all pending lawsuits and established universal acknowledgment of the validity of both companies' patents. As a result, Affymetrix gained access to a commercial license of Hyseq' array related technologies, and Hyseq gained access to an internal use license of Affymetrix technology for its pharmaceutical business. Both companies also formed a collaborative venture known as Nmer, Inc. to develop the sequencing by hybridization applications using both Hyseq's and Affymetrix's technologies.

Oxford Gene Technologies and Affymetrix: Oxford Gene Technologies (OGT) (Oxford, UK), a company established by Edwin Southern, issued two fundamental patents regarding the manufacture and use of DNA microarrays, US5700637 and US6054270. Southern is globally recognized for his key insight over 25 years ago that labeled nucleic acid molecules can be used to interrogate genetic material, to determine DNA sequences. Both the US5700637 and US6054270 patents claim using a solid support, such as a glass plate or film,

11

containing an array of oligonucleotides to identify DNA sequences, under hybridization conditions where discriminations can be made between matched and mismatched oligonucleotide probes. The US5700637 patent claims the synthesis of oligonucleotide arrays in situ whereas the US6054270 patent covers attaching synthesized oligonucleotides to solid supports. Litigation proceedings between OGT and Affymetrix began on June 4, 1999, when OGT filed against Affymetrix for infringement of the US5700637 patent. At issue was whether Affymetrix secured a valid license to use the patent. In 1991, OGT (then Isis Innovations) issued an exclusive license to Beckman Coulter to commercialize Southern's technology. The provisions of this agreement included that the license could be shared if Beckman Coulter entered a consortium with a collaborator. Alternatively, the license could be transferred if Beckman Coulter sold the part of the company that was developing this technology. In May 1998, Affymetrix attempted to acquire Southern's license by entering into a consortium with Beckman Coulter to commercialize the technology, and in July 1999 they entered into an agreement with Beckman Coulter to purchase the patent license. The Court of Appeals in the UK ruled in November 2000 that Affymetrix possessed a valid license to the OGT microarray technology when it purchased Beckman Coulter's microarray business.

A separate OGT lawsuit at the High Court in the UK for revocation of Affymetrix EP 619,321 patent, which also included density claims, was settled in March 2001 and Affymetrix agreed to license OGT's technology.

In December 2002, OGT initiated litigation proceedings against several companies, including Nanogen (San Diego, CA, USA), BD Biosciences Clontech (Palo Alto, CA, USA), Genomic Solutions (Ann Arbor, MI, USA), Perkin Elmer (Boston, MA, USA), Axon Instruments, Inc. (Union City, CA, USA) and Biodiscovery (Marina Del Rey, CA, USA) for infringement of the US6054270 patent. Currently, several parties are challenging the validity of the US5700637 patent at the EPO. Thus, the lawsuits for the most basic filings are not settled completely.

Incyte versus Affymetrix: Spotted microarrays represent an important format of this technology. This system was developed by Patrick Brown and colleagues at Stanford University, CA, USA and is claimed in the US5807522 and US6110426 patents. This method involves dispensing a known volume at each selected array position by tapping a capillary dispenser on a support in order to deposit a defined volume of liquid. This patent also claims an automated apparatus for forming a microarray. The US5807522 patent demonstrates the manufacture of arrays consisting of genomic DNA fragments and cDNAs in addition to multiplexed colorimetric hybridizations. Synteni, founded in 1994, initially commercialized this technology until it became the microarray division of Incyte when it was acquired in 1997.

Incyte acquired from Stanford University the patents that cover basic methodologies of preparing samples for hybridization experiments. Patents US5716785 and US5891636 claim methods of amplifying RNA from cDNA where cDNA synthesis is primed with a promoter sequence that initiates transcription fused with an oligo dT sequence. This technique has greatly improved the sensitivity of microarray experimentation, allowing expression profiling with very small amounts of RNA, such as that typically available from tissue biopsy and laser-capture material.

Affymetrix filed a lawsuit on January 6 1998 against Incyte for infringement of the US5445934 patent, on the basis that this patent covers arrays of 1000 or more oligonucleotides in a square centimeter. Regarding this patent, Incyte argued that the patent described the word 'oligonucleotides' to mean single-stranded polymers synthesized monomer by monomer. Affymetrix contended that oligonucleotides mean small polymeric stretches of nucleotides ranging from 2 to ~100 nucleotides. The 9th Circuit District Court eventually rejected Incyte's argument by looking at the plain definition of oligonucleotides, which is small polymers of nucleotides. The court held that Incyte's interpretation of the word oligonucleotide is unconventional and that Incyte failed to support any portion of the specification or prosecution as evidence to show that the patentee intended the word oligonucleotide to have this unconventional meaning.

The issue regarding the interpretation of the phrase 'discrete known regions' was also a topic in this litigation. Incyte argued that this meant a localized area on a surface intended to be activated for synthesis of a polymer. Affymetrix articulated this to mean a localized area intended to be 'activated for formation of a polymer'. The patentee went further to state that formation encompasses both synthesis on a surface and immobilization of pre-existing polymers on the surface. Affymetrix's definition was problematic because in the patent there was no explicit definition for the term, 'activated'.

The court used the patent's specification to determine that this word means an energy source adapted to cause a chemical group active for polymer formation. Further, the court also found that 'formation' means immobilization of polymers and therefore rejected Incyte's contention that 'discrete known regions' should be limited to what was described in the patentee's specification.

At the time of the initiation of this lawsuit, Incyte/Synteni were engaged in an interference proceeding to invalidate the US5744305 and US5800992 patents. In the US5744305 patent, Affymetrix in one of its claims replaced the word 'oligonucleotides' with the word 'polynucleotides' which was intended to mean a polymer of nucleotides of length two or more. Incyte argued that 'polynucleotides' means a strand of DNA that is generally naturally-occurring and longer than an oligonucleotide. The court rejected Incyte's assertion by using the plain meaning of the word polynucleotide that encompasses short oligonucleotides as well as longer stretches of DNA.

The US5800992 patent concerns the usage of adding mixtures of nucleic acids derived from two different cell types with different labels into an array in order to measure differential gene expression. In September 1998, Affymetrix sued Incyte for infringement of this patent. Incyte argued that this patent did not cover using cDNA as a means to measure differential expression since cDNA is made outside the cell, and the patent is limited to using nucleic acids from the cell. The court agreed with this argument. This was an important ruling since nucleic acids are generally converted into cDNA before microarray processing. In August 2000, Incyte sued Affymetrix for infringement of the US5716785 and US5891636 patents regarding the Eberwine sample preparation technique for microarrays that use the cDNA synthesis technique.

In December 2001, the companies settled all pending litigation and each agreed to certain non-exclusive, royalty-bearing licenses and internal use licenses under their respective IP portfolios. Affymetrix paid Incyte US$4.5 million for use of the RNA amplification technology covered by the US5716785 and US5891636 patents.

Incyte used the enabling requirement under Section 112 to argue that no infringement had occurred of the US5445934, US5744305 and US5800992 patents because these patents failed to demonstrate methods for manufacturing cDNA arrays. This is possibly why Affymetrix was unable to establish that Incyte was liable for infringement, although this issue was not actually decided in a courtroom as the parties opted to settle their litigation. Incyte has continued to enforce the US5807522 patent for methods of manufacturing cDNA arrays, and in November 2001, the company filed a lawsuit against Invitrogen (Carlsbad, CA, USA) for infringement of this patent as well as the US5716785 and US5891636 sample preparation patents.

In 2002, Incyte terminated their microarray business, selling all of their microarray technology assets to Quark Biotech (Cleveland, OH, USA). However, Affymetrix is recognized as the patent holder for microarrays printed at various densities. Affymetrix has reported in its press releases that NEN Life Science Products, MWG-Biotech AG and Genomic Solutions are three of 25 companies that have obtained a license to make and use spotted microarrays.

In the case of Genomic Solutions, securing a license from Affymetrix to manufacture spotted microarrays did not protect the company from potential liability. OGT sued the company for infringement of the US6054270 patent for spotting synthesized oligonucleotides on arrays. Thus, in order to understand which companies to secure licenses from, it is very important to examine the patents granted and ensure that the specifications include all of the methods covering the type of technology that will be utilized.

Outlook–microarrays and the future: The boundaries of IP space will be continually challenged as this technology progresses. Many different approaches to manufacturing and processing samples using microarrays are being developed. Standard phosphoramidite DNA oligonucleotide synthesis has been performed in situ using an inkjet printing method. Combimatrix (Mukilteo, WA, USA) have established a solid-phase oligonucleotide synthesis system using a method that electrochemically places monomers at specific locations on substrates. Alternatives to conventional photolithography with chromium masks are evolving. One alternative to photolithographic masks for in situ oligonucelotide synthesis is the use of a digital micromirror device (DMD or DLP) which is employed within the technologies of Nimblegen (Madison, WI, USA), Febit (Mannheim, Germany) and Xeotron (Houston, TX, USA). Nimblegen have described a technology for a centralized production facility that can synthesize microarrays containing 195,000 features using a digital light processor that creates digital masks to synthesize specific polymers.

Whereas Nimblegen and Xeotron have used a conventional microscope slide as the reaction carrier, Febit has introduced an integrated benchtop instrument that generates microarrays within three-dimensional microstructures (microchannels). Xeotron have recently switched from microscope slides to microstructures. Microarray apparatuses containing microfludic structures are also being developed. Microfluidicbased systems have been established that detect hybridization events using electrochemical methods such as voltammetry, amperometry and conductivity, that will alleviate the need for target sample labeling. Metrigenix (Gaithersburg, MD, USA) have developed a Flow-Thru Chip™ four-dimensional microarray for high-throughput assays. Unlike conventional arrays the molecular interactions occur within the three-dimensional volumes of ordered microchannels rather than on two-dimensional flat surfaces. A fourth dimension is created by microchannels, which connect the upper and lower faces of the chip, allowing fluid to flow through the chip.

In addition to the evolving technical approaches of DNA microarray systems, new applications for microarrays are being developed. Microarrays are being employed as gene delvery vectors that transfect cell monolayers grown on the array surface. Furthermore, tissue microarrays are enabling researchers to perform tissue analysis in a high-throughput, parallel manner. Microarrays are currently poised to enter the clinical arena. Automated chip platforms permitting multiplexed assays such as the INFINIT ™System from Autogenomics (Carlsbad, CA, USA) will facilitate improved genetic testing. Their emergence and success in the clinical laboratory will be dependent on their ability to adapt to the rigorous environment of routine usage.

The challenge for technology platform companies is to effectively establish and communicate the degree of their respective exclusivities in order to remain competitive in this continually developing market place. Close attention to detail regarding this IP space is very important in this highly competitive field. Utilizing professional consultation and resources is highly recommended when developing or working with this type of technology.

Business Strategies

What is interesting about biochips is that the field is changing fast. Biochips are becoming more diverse and customers are appearing in other world regions such as China and India. The products from the biochip market are being used to help accelerate the research processes and capabilities of bio-pharmaceutical drug discovery and basic academic bioresearch.

During the past ten years, major pharma companies have seen their R&D costs explode as their actual productivity has declined. Drug companies have looked outside of their labs to find drug products faster. As a result, pharma companies have increasingly become motivated research and financial partners with many biotechnology companies. At the same time, pharma companies also want to become customers and partners of biochip companies because these companies have technologies that might help them become more productive faster.

New business models are being created. The ideas of pharmacogenomics and targeted therapies have received positive reception from government regulators. Roche and Affymetrix were the first to receive FDA approval for a biochip-based molecular diagnostic test and instrument system. Basic academic research is helping too. Accelerating the drug making process requires modern industrialization of R&D, by using genomics and proteomics technologies such as biochips that the biotechnology companies provide.

11

Common strategies employed by microarray companies

- Markey survey.
- Identifying of the core areas for business.
- Investment in R&D in developing novel technologies.
- Patenting of novel technologies.
- Acquiring patents.
- Development of business models.
- Acquiring startups.
- Development of marketing strategies for sales of the products.

Markets for Microarray

The global market for microarray: DNA microarrays first appeared on the life science scene in February 1991. It was then that the journal Science published the paper "Light-directed, spatially addressable parallel chemical synthesis". Having detailed the technology for synthesizing oligonucleotides on a tiny glass chip, Stephen PA Fodor, Ph.D., the paper's first author, went on to create the first DNA microarray, which he eventually named the GeneChip. In 1992, Dr. Fodor founded Affymetrix. The company launched the first commercial DNA microarray in 1994. In the fiscal year ending in December 31, 1995, the company brought in revenues of $4.6 million. During the next fiscal year, revenues were $12.0 million. Nine years later, during the fiscal year ending December 31, 2005, Affymetrix's revenues are at nearly $368 million.

Since its inception , the use of microarrays has spread rapidly throughout the research community. The recent technological advancements in microarray analysis have revolutionized understanding fundamental aspects of the growth and development as well as the molecular mechanisms underlying causes of many human diseases. Microarray technology helps in understanding the structure-function relationship, thereby heralding the era of New Biology. Researchers who carry out microarray experiments ask big questions and the results generate large quantities of data. Bioinformatics power therefore needs to step up to the challenges that have been created by microarray technology.

This revenue growth has been brought about because of the fact that nucleic acid microarrays have found many uses. Basic research labs use microarrays to examine gene expression and single nucleotide polymorphisms, characterize genotypes and compare whole genomes. While drug discoverery labs have found microarrays efficient tools for lead optimization, toxicity prediction, screening patients for clinical trials and monitoring their progress.

Right place, right time: The arrival of the DNA microarray could not have come at a more optimal time. During the late 1980s, gene sequencing efforts were beginning to gain steam with improved methods and the introduction of the first automated sequencer by Applied Biosystems Group (Foster City, Calif.), which is now part of Applera Corporation (Norwalk, Conn.). By the early 1990s, researchers began examining the sequenced genes and their functions, thus launching the genomics era. Microarrays liberated researchers from the tedious process of studying gene sequences individually. By providing the capability to simultaneously screen thousands of oligonucleotides, microarrays quickly became a standard tool for studying gene expression.

With such potential, many companies have joined the fray with their own brand of nucleic acid microarrays. Now, leading companies include Agilent Technologies Inc (Palo Alto, Calif.), GE Healthcare Bio-Sciences (Little Chalfont, UK), Illumina Inc (San Diego, Calif.), according to Cambridge Healthtech Advisors. However, the combined sales of these three companies equate to just over a third of Affymetrix's sales. In fact, the microarray giant accounts for more than half of all microarray units sold. Affymetrix also leads in instruments for microarray readers and other related accessories. Whereas Agilent reports placing its microarray instruments with 400 customers, Affymetrix has placed more than 1,000 instruments.

The microarray giant also dominates in the areas of licensing, marketing and distribution deals, according to Kalorama Information. In regards to R&D alliances, Affymetrix shares the distinction of being most active with Agilent and CombiMatrix Corp (Mukilteo, Wash).

NimbleGen Systems Inc (Madison, Wisc.) joined Affymetrix and Agilent as having catalogs with the broadest selection of arrays. These three companies also lead the pack in terms of custom arrays, which are slated for major growth. In 2004, Affymetrix moved to secure its position in custom arrays by partnering with NimbleGen, whose photolithographic Maskless Array Synthesis technology enables flexibility, which is advantageous in producing custom arrays. During the same year, Invitrogen Corp (Carlsbad, Calif.) entered the market with its acquisition of Xeotron Corp (Houston, Texas), which also specialized in custom arrays.

Rewarding returns: Such alliances and acquisitions demonstrate the vitality of the nucleic acid microarray market. In 2005, DNA microarrays brought in revenues of $446.8 million, according to Frost & Sullivan. The global market for microarrays with an estimated revenue of $ 5.3 Billion currently(2009) is anticipated to grow to a projected revenue of $ 10 Billion by the year 2012. Affymetrix microarray is considered to be the Gold Standard and is the market leader with over 2000 platforms. Microarray technology is dominated by aftermarket sales comprising of arrays and reagents. Aftermarket is the fastest growing microarray market segment. Forty-five per cent of the microarray users are from academia.

As prices have dropped, users have been relying on commercial arrays, rather than homebrew products. Eighty per cent of the microarray users are from the research and development. The majority of the R & D projects take place within pharmaceutical and biotechnology industries, followed by academia and government labs. North America is the largest consumer of microarray and accounts for 45% of the global microarray business.

Great prospects, grand plans: The chip format also faces competition from the popularity of bead-based assays, with Illumina, Luminex Corp (Austin, Texas) and Quantum Dot Corp (Hayward, Calif.) as notable players. Microarrays based on 96-well microplates have also gained a large following. Companies will continue to offer arrays representing certain disease areas and cell or tissue types for an increasing number of organisms. Microarrays hold great potential for penetrating the $20 billion in vitro diagnostics industry and supporting the burgeoning area of pharmacogenomics, which aims to not only improve diagnoses, but also treatment decisions. But these possibilities await efforts to characterize and validate genomic signatures for specific conditions and susceptibilities. Then, such approaches face many hurdles of the FDA before reaching the market place.

Indian market: It was in early 2000 when Indian Institute of Science, Bangalore established the first spotted microarray facility in India. Since then the microarray systems market in India has witnessed some level of growth but the major technological advancements which came up in the last couple of years has given the real push. There are many suppliers of different microarray technologies–oligo-based photolithography(Affymetrix), oligo-based inkjet (Agilent), Bead based(Illumina), Maldi-based (Sequenom), etc. The traditional cDNA-based microarray systems are also available from Molecular Devices Corporation(Axon), Perkin Elmer and Genomic Solutions.

Approximately there are over 75 microarray scanners and systems installed in the country so far. Even relatively late entrants into the Indian market such as Affymetrix have been able to quickly build up over a dozen GeneChip platforms within 3 years. Indian microarray business for 2009 is estimated to be around $ 9 million and is expected to grow to $ 15 million by the year 2012.

The impetus is provided by the Department of Biotechnology which is the nodal funding agency for biotech research in India. Centers of Genomic Excellence are being set up for effective utilization of such technologies to Indian needs. Many scientists are using microarrays for their analysis of gene expression and genotyping. Microarray systems are being utilized in various research institutes, universities, medical research labs, agricultural research labs, pharma industries and service providers. Many customers have graduated themselves from cDNA to oligo-based technology. Most of the cDNA slides are imported and some are spotted within the country. Consortia for Human SNP, Rice genome, etc exist in the country. Many large projects are being undertaken for various life style diseases, both for the analysis of gene expression and genotyping. Many publications have appeared in international journals. Many research communities are interested in custom arrays for agriculture and healthcare.

Future of Microarray

Biochips will evolve both in terms of their physical characteristics and also in terms of the assays they carry out. Biochips will consistently grow smaller and more powerful with each generation of biochip created. Additionally, with the development of specialty biochips, based from various organic materials can lead to new developments and utilizations. One encouraging development is protein based biochips. The so-called patterning of proteins on various chip substrates is the focus of intense research, with examples including development of three-dimensional patterning vs. nano-patterning on single layers. These biochips would be used to array protein substrates for drug lead screening, antibodies for diagnostic purposes, where the biochip then is called as biosensor, enzymes for catalytic reaction analysis and other applications.

A broad "Technology Platform", biochip technology will redefine "genetic diagnostics, because of their reproducibility, low cost and speed. They also promise to open new areas of drug screening development once proteins and other DNA molecules are successfully deposited, and their reactions read by detectors, onto chips". However, much of the technology that is used to design, produce and analyze the results of biochips is labor intensive and expensive.

Better methods of manufacture and detection are the paramount in order of mainstream adoption of biochip biotechnology. Currently new methods are being developed to increase efficiency without compromising product quality. Techniques include direct detection of hybridization by electric current and standardization of biochip technologies for mainstream compatibility. The industry has applied biochip technology to commercial use. The market leader Affymetrix currently is the only provider of commercial use biochip technology. However, many companies, sensing long-term profit potential, have entered into biochip technology as the technology becomes more accessible. Although limited in profit and scope, biochip technology has a relevant future in medical, forensic, and commercial applications.

Index

Reader's Notes

Reader's Notes

Reader's Notes